Palgrave Historical Studies in the Criminal Corpse and its Afterlife

Series Editors
Owen Davies
School of Humanities
University of Hertfordshire
Hatfield, UK

Elizabeth T. Hurren
School of Historical Studies
University of Leicester
Leicester, UK

Sarah Tarlow
History and Archaeology
University of Leicester
Leicester, UK

This limited, finite series is based on the substantive outputs from a major, multi-disciplinary research project funded by the Wellcome Trust, investigating the meanings, treatment, and uses of the criminal corpse in Britain. It is a vehicle for methodological and substantive advances in approaches to the wider history of the body. Focussing on the period between the late seventeenth and the mid-nineteenth centuries as a crucial period in the formation and transformation of beliefs about the body, the series explores how the criminal body had a prominent presence in popular culture as well as science, civic life and medico-legal activity. It is historically significant as the site of overlapping and sometimes contradictory understandings between scientific anatomy, criminal justice, popular medicine, and social geography.

More information about this series at
http://www.springer.com/series/14694

Sarah Tarlow

The Golden and Ghoulish Age of the Gibbet in Britain

palgrave
macmillan

Sarah Tarlow
University of Leicester
Leicester, UK

Palgrave Historical Studies in the Criminal Corpse and its Afterlife
ISBN 978-1-137-60088-2   ISBN 978-1-137-60089-9   (eBook)
DOI 10.1057/978-1-137-60089-9

Library of Congress Control Number: 2017951552

© The Editor(s) (if applicable) and The Author(s) 2017. This book is an open access publication.
The author(s) has/have asserted their right(s) to be identified as the author(s) of this work in accordance with the Copyright, Designs and Patents Act 1988.
Open Access This book is licensed under the terms of the Creative Commons Attribution 4.0 International License (http://creativecommons.org/licenses/by/4.0/), which permits use, sharing, adaptation, distribution and reproduction in any medium or format, as long as you give appropriate credit to the original author(s) and the source, provide a link to the Creative Commons license and indicate if changes were made.
The images or other third party material in this book are included in the book's Creative Commons license, unless indicated otherwise in a credit line to the material. If material is not included in the book's Creative Commons license and your intended use is not permitted by statutory regulation or exceeds the permitted use, you will need to obtain permission directly from the copyright holder.
The use of general descriptive names, registered names, trademarks, service marks, etc. in this publication does not imply, even in the absence of a specific statement, that such names are exempt from the relevant protective laws and regulations and therefore free for general use.
The publisher, the authors and the editors are safe to assume that the advice and information in this book are believed to be true and accurate at the date of publication. Neither the publisher nor the authors or the editors give a warranty, express or implied, with respect to the material contained herein or for any errors or omissions that may have been made. The publisher remains neutral with regard to jurisdictional claims in published maps and institutional affiliations.

Cover illustration: © Melisa Hasan

Printed on acid-free paper

This Palgrave Macmillan imprint is published by Springer Nature
The registered company is Macmillan Publishers Ltd.
The registered company address is: The Campus, 4 Crinan Street, London, N1 9XW, United Kingdom

Acknowledgements

The book is based on research carried out as part of the research programme "Harnessing the Power of the Criminal Corpse", funded by the Wellcome Trust (grant WT095904AIA). I am very grateful to the Wellcome Trust for their support and assistance—financial and practical. I also thank the other participants in the programme: Elizabeth Hurren, Peter King, Richard Ward, Zoë Dyndor, Shane McCorristine, Owen Davies, Francesca Matteoni, Floris Tomasini, Rachel Bennett and Emma Battell Lowman. Many thanks to Emily Russell and Rowan Milligan at Palgrave and to Steve Poole for his extremely helpful comments on the first version of this book. Thanks to Adam Barker for invaluable assistance with the index. Most images are my own, unless otherwise specified.

Contents

1	Some Further Terror and Peculiar Mark of Infamy	1
2	How to Hang in Chains: How, Where and When Eighteenth-Century Sheriffs Organised a Gibbeting	33
3	The Afterlife of the Gibbet	79
4	Conclusions: Why Gibbet Anyone?	101
Appendix 1: All Cases of Hanging in Chains		119
Appendix 2: Maps, 1752–1834		135
Concept Index		145
Historical Publications Index		147
Name Index		149
Place Index		153

List of Figures

Fig. 1.1	Number of gibbetings per decade in England and Wales, 1700–1832	28
Fig. 2.1	St Peter's rock, Derbyshire, where Anthony Lingard was hung in chains in 1815	41
Fig. 2.2	Felton's obelisk in Portsmouth	42
Fig. 2.3	Road sign, Gibbet Hill Lane, Scrooby	47
Fig. 2.4	Interval in days between the first day of the assizes during which a criminal was convicted and the date of his execution	63
Fig. 2.5	A Thames pirate	65
Fig. 2.6	Tull or Hawkins's leg iron, courtesy of Reading Museums	66
Fig. 2.7	Some different styles of gibbet: a: John Breeds (Rye, 1743, now in Rye town hall); b: John Keal (Louth, 1731, now in Louth Museum); c: possibly 'Jack the Painter' (Portsmouth, 1777, now in Winchester Museum); James Cook (Leicester, 1832, replica now in Leicester Guildhall)	70
Fig. 2.8	Multiple punches holes on John Keal's gibbet	71
Fig. 2.9	Headpiece of John Breeds's gibbet with large skull fragment remaining	71
Fig. 2.10	Artistic representation of a gibbet with carrion birds. Vignette from Thomas Bewick's *British Birds* (1804)	72
Fig. 3.1	'Willow biter' and rhyme, drawn and recorded in the commonplace book of Edwin Jarvis of Doddington Hall, Lincs., courtesy of Claire Birch	88

Fig. 3.2	'Noose and Gibbet' pub, Sheffield	90
Fig. 3.3	Eugene Aram's skull	96
Fig. 3.4	Gustave Doré's engraving of Eugene Aram	98

List of Maps

Map 1a	1752–1760	136
Map 1b	1761–1770	137
Map 1c	1771–1780	138
Map 1d	1781–1790	139
Map 1e	1791–1800	140
Map 1f	1801–1810	141
Map 1g	1811–1820	142
Map 1h	1821–1830	143
Map 1i	1831–1834	144

List of Tables

Table 1.1	Numbers hung in chains under the Murder Act	10
Table 1.2	Crimes punished by hanging in chains, 1752–1832	22
Table 2.1	The frequency of gibbetings by county and decade through England and Wales	37
Table 2.2	Admiralty Court convictions resulting in hanging in chains	51
Table 2.3	Surviving gibbet cages	64

CHAPTER 1

Some Further Terror and Peculiar Mark of Infamy

Abstract The story of Tom Otter, a murderer who was executed and gibbeted in 1806, has many striking features. Not least, this form of brutal and bodily post-mortem punishment seems rather anachronistic during a period often described in terms of increasing gentility and humanity. It took place within the legal context of the Murder Act (1752), which specified that the bodies of murderers had to be either dissected or hung in chains. Other aggravated death penalties were applied to those convicted of treason and suicide. A number of common misconceptions about the gibbet need to be corrected.

Keywords Tom Otter · Murder act · Suicide · Treason · Post-mortem punishment

Tom Otter

Tom Otter was not what he seemed. In fact, when he murdered his second wife on their wedding day in 1805, he wasn't even called Tom Otter. A bigamist, a murderer, a corpse and a ghost, Tom Otter was as unreliable as the numerous stories that were told about him from the time of his arrest to the present day. These included the rumour that he had murdered his baby (untrue: his wife was pregnant when he killed her, but had not given birth), that somehow contrived to murder another man after his own death by causing his gibbet cage to fall and crush him (also untrue), and that every year on the anniversary of his wife's murder, his ghost would cause the hedge stake with which the bloody deed was committed to appear, covered in gore, at the scene of the crime (a great story, but based on a mid-nineteenth-century fiction).

What we do know about Tom Otter is less sensational and more grim. Thomas Otter was born in the Nottinghamshire village of Treswell in 1782 and married Martha Rawlinson there in 1804, the same year that their daughter was christened at Hockerton. However, the very next year, he found navvying work on the canals of Lincoln. He was at that time calling himself Thomas Temporel, his mother's maiden name and the name under which he was soon to stand trial for murder. While in Lincolnshire, he seems to have quickly forgotten his wife and child in Treswell and taken up with a local girl called Mary Kirkham who, in due course, also became pregnant. To avoid the problem of illegitimacy and the need to support unmarried mothers and bastard children on parish relief, Otter/Temporel was compelled to marry Mary Kirkham on 3 November 1805, when she was about eight months pregnant. The South Hykeham parish register records that their marriage was witnessed by William and John Shuttleworth, the Overseers of the Poor for that parish. This is evidence that their wedding was a so-called "knobstick" marriage—like a "shotgun wedding", this was a forced union intended to compel fathers to take responsibility for their own illegitimate children. Instead of the bride's angry father being the driving force, representatives of the local parish who would have to provide for unsupported women and children were the principal enforcers of knobstick unions. But Tom and Mary's marriage was very short-lived. Later that very same day when the newly married couple were on their way back to Doddington where he lived, Thomas attacked Mary with a hedge stake and killed her at a place called Drinsey Nook.[1]

Tom was arrested the following day and brought to Lincoln castle. Mary's body was taken to the local inn (the Sun Inn in Saxilby) for post-mortem examination. Her body was subsequently buried in the north-east corner of Saxilby churchyard. Otter's guilt was never really in doubt and at his trial, during the March assizes of 1806, he was sentenced to

---

[1] This history of Tom Otter is much indebted to the excellent work carried out by the Saxilby and District History Group and published at http://www.saxilbyhistory.org/

death and dissection in accord with the 1752 Murder Act. Before the judge left town, the post-mortem part of the sentence was changed to hanging in chains.

Accordingly on March 14, Tom Otter was hanged at Lincoln gaol. After his death, his body was encased in a gibbet cage for which he had been measured before his execution—an experience upon which "all his fortitude appeared to forsake him".[2] His body was then transported to Saxilby and the gibbet cage was hung up on a pole thrity feet high on Saxilby Moor, about 100 yards from the place where Mary's body had been found. A huge crowd gathered to see the body being hung on the gibbet and for many days afterwards the scene was, according to an eye-witness "just like a fair".[3] Another man remembered his father's account: "For several days after the event, the vicinity of the gibbet resembled a country fair with drinking booths, ballad singers, Gypsy fiddlers, and fortune-tellers".[4]

This was not, however, the end of Tom Otter's story. Not only was his gibbet thronged with visitors during the early days, it remained suspended for more than forty years while his remains gradually decayed and fell away. Only a violent storm in 1850 finally brought the gibbet cage down. On that occasion, the lord of the manor, Edwin George Jarvis, recorded in his notebook that he managed to acquire the headpiece, though "the gypsies made off with nearly all the remains",[5] presumably for their value as scrap metal. The headpiece is still kept at Doddington Hall, Jarvis's home and now home to his descendant, Claire Birch.

Given its prominence in the landscape and the memorable circumstances of its erection—one can be fairly sure that the murder of Mary Kirkham and the subsequent execution and gibbeting of Tom Otter must have been among the most dramatic and thrilling—if disturbing— things that ever happened in Saxilby, it is not surprising that the gibbet left enduring traces in the landscape. Though the exact location of the gibbet is not marked, the road on which stands is called Tom Otter's

---

[2] *The Lincoln, Rutland and Stamford Mercury*, 21 March 1806.

[3] This quotation, and much of the story, is taken from of Edwin George Jarvis's unpublished commonplace book, which is in the possession of Claire Birch of Doddington Hall, Lincs.

[4] George Hall (1900) *The Gypsy's Parson* (London: Marston and Co), p. 17.

[5] Commonplace book of Edwin Jarvis.

Lane, which leads to Tom Otter's Bridge. Nearby are Gibbet Woods and Gibbetwood Farm. Gibbet Lane cottages lie a little way to the southeast.

As well as writing his name and fate permanently into the landscape around the scene of his crime, Tom Otter persists in some pieces of local folklore. The first concerns the malevolent spirit of Otter himself. Legends—now perpetuated mostly on the internet—tell how the weight of Otter's gibbet cage was so great that it fell twice from its post, the second time killing a man who had earlier taunted Otter. Then there is the story of how every year, on the anniversary of Mary Kirkham's murder, the hedge stake with which Otter committed the deed was found to be missing from the wall of the Peeweet (now Pyewipe) Inn and turned up instead in the field where she died, covered in blood. Even when a group of men decided to stay up and keep watch, they all mysteriously fell asleep at the same time and on waking found that the hedge stake had gone to the field once more. In the end, the story says, the hedge stake could be stilled only when the Bishop of Lincoln burned it outside the Cathedral. Another tale is that the Sun Inn, where Mary's body was brought for inquest, is haunted by the ghost cries of Tom Otter's baby.

Interestingly, all of these tales can be traced to a story published in the *Lincoln Times* in 1859 by Thomas Miller.[6] The Lincolnshire Record Office holds the covering letter that Miller wrote when sending his Tom Otter story to the *Lincoln Times*, from which it is very clear that the story is meant to be fiction, with only a small core of historical fact. Nevertheless, the ghosts of Drinsey Nook are a regular fixture in the investigations of paranormal interest groups and Lincolnshire ghost tours.

Post-mortem Punishment

Tom Otter's tale has many commonalities with the later parts of other criminal histories of the long eighteenth century. For the historian or archaeologist, it also raises a number of interesting questions. What were the purpose and meaning of the rather repulsive practice of hanging in chains? What did it actually entail? What effect did it have on the criminal, on the justice system and on the huge crowds who witnessed the event and the even larger numbers who eagerly consumed journalistic or

---

[6] Maureen James 2011. http://tellinghistory.co.uk/content/additional-information-not-included-lincolnshire-folk-tales-maureen-james-published-history.

fictional accounts of gibbets and their inhabitants? What kind of mental and physical legacy was left by the gibbets which formerly stood by roadsides and on commons all over England? This short volume picks up where most crime historians leave off, when the lifeless (or apparently lifeless) body is hanging from the execution scaffold, and follows the corpse into its gibbet irons where it might remain for many decades. This exploration makes use of archaeological, landscape, folkloric and literary evidence where relevant, but most of its data comes from historical newspaper and archival sources. In particular, it makes use of the invaluable "sheriffs' cravings", which are the expense claims submitted by county sheriffs, usefully detailing the practical elements of carrying out sentences, now stored in the National Archives at Kew.

Principally we are concerned here with the period from the Murder Act of the mid-eighteenth century to 1832, when the last gibbeting took place. Most examples are English and although I will be drawing in occasional examples from the other countries of the British Isles, there is no attempt to look at the global history of hanging in chains. This chapter looks at the legal background to the punishment and briefly considers other forms of post-mortem punishment before asking the question, "Who was hung in chains, and what were the circumstances that made hanging in chains, rather than another means of post-mortem punishment, the appropriate choice?"

Hanging in Chains Before the Murder Act

Hanging in chains predates the 1752 Murder Act and was a widely used punishment in the earlier eighteenth century and the seventeenth century. The same is also true of dissection, both punishments being part of the discretionary repertoire of the judge. However, the genealogies of the two treatments are different. The use of criminal corpses for anatomical dissection was driven principally by the needs of the anatomists. As Richardson has discussed, the earliest regular supply of cadavers for dissection was the result of legislation in the time of Henry VIII specifying that the bodies of four executed felons be supplied to the Barber Surgeons each year. By contrast, hanging in chains is a punishment more related to the bloodthirsty retributive punishments of the late medieval and early modern periods. The display of bodies—or more often of body parts, especially the head—was a common element of punishment for serious crimes such as murder or treason before the eighteenth century and was carried out in England as part of the sentence for treason as late as 1745–1746 after the Jacobite rebellion.[7] The display of body parts in the medieval and early modern periods was particularly associated with crimes against the State or the political order. Body parts were typically displayed above city walls and gates or on prominent public buildings. The particular geographical specificity of hanging in chains as a post-execution punishment which is tied to the scene of crime was an effective way of perpetuating the memory of an atrocity. This goes some way to explaining its popularity in the punishment of aggravated highway robbery, and the tradition of hanging in chains those who have committed murder on the highway seems to have been established during the seventeenth century. Thomas Randall was punished this way for murder and robbery on the highway in 1696 and added to his spectacular death by dressing all in white for his execution.[8]

THE MURDER ACT

Tom Otter's sentence for murder was not only execution—which was well established as the usual punishment for such a crime—but also the stipulation that after death his body was to be "hung in chains". In the early nineteenth century, the sentencing of Otter's crime was determined by the Murder Act. The 1751 act (which came into force in 1752 and so is often attributed to that year) was called "An Act for Better Preventing the Horrid Crime of Murder" and was known generally as the Murder Act. It was largely superseded by the Anatomy Act of 1832 and was formally abolished in 1834.

The punishment for murder in the middle of the eighteenth century, as it had been for many centuries before, was death. However, by that time, the number of crimes for which the penalty was death was more than 220[9], compared with around 50 capital offences in 1688.[10] When

---

[7] V.A.C. Gatrell (1994) *The Hanging Tree: execution and the English people 1770–1868* (Oxford: Oxford University Press), p. 317.

[8] *Post Man and the Historical Account*, 114, 30 January 1696.

[9] D. Levinson (2002) *Encyclopedia of Crime and Punishment, vol. 1* (Thousand Oaks, CA: Sage), p. 153.

[10] H. Potter (1993) *Hanging in judgement: religion and the death penalty in England from the bloody code to abolition* (Ann Arbor: SMC Publishing), p. 4.

you could, in theory, be hanged for poaching rabbits or going out after dark with a blackened face, the issue of distinguishing the most serious crimes became a problem.[11] Peter King has studied the extensive eighteenth-century public debate about what would constitute an appropriate and effective punitive response to serious and violent crime. Suggestions included ways of exacerbating the pain of execution through, for example, breaking on a wheel, as was widely practised elsewhere in Europe, or torturing to death. Some commentators advocated the use of some kind of *lex talionis*, which follows the principle that punishment should mimic whatever was inflicted on the victim of a crime. Thus, murder by drowning would be punished by drowning the perpetrator; serious assaults might be punished by inflicting a similar wound on the criminal before his or her execution.[12] Alternatively, the punishment of execution could be augmented by spreading the subject of punishment to include the criminal's family. Finally, the punishment might be extended past the point of death by causing an element of post-execution violence or humiliation to be enacted on the dead body of the criminal. In the case of suicides, men who had escaped the dock before death were subject to all those forms of post-mortem punishment.[13] A long period of debate about exacerbated forms of punishment preceded the introduction of the 1752 bill, and indeed the extension of post-execution punishment to crimes other than murder continued to be advocated during the later eighteenth century. In particular, serious attempts

---

[11] In fact, as historians have shown, during the period of the so-called "Bloody Code", the discretion of the judges and the reluctance of the juries meant that discretionary death sentences for property crime were often avoided or reprieved. This has led King and Ward to suggest that the long eighteenth century in England was in fact the period of the Unbloody Code. See P. King (2000) *Crime, justice and discretion in England 1740–1820* (Oxford: Oxford University Press); P. King and R. Ward (2016) 'Rethinking the Bloody Code in Eighteenth-Centre Britain: Capital Punishment at the Centre and on the Periphery' *Past and Present* (2016); J. Beattie (1986) *Crime and the Courts in England 1600–1800* (Princeton: Princeton University Press).

[12] Peter King (forthcoming) *Punishing the Criminal Corpse 1700–1840: aggravated forms of the death penalty in England* (Basingstoke: Palgrave).

[13] Rab Houston (2011) *Punishing the Dead: suicide, lordship and community in Britain 1500–1830.* (Oxford: Oxford University Press), p. 203; Robert Halliday (1997) 'Criminal graves and rural crossroads' *British Archaeology* 25 (June 1997); M. MacDonald and T. Murphy (1990) *Sleepless souls: suicide in early modern England* (Oxford: Clarendon Press).

were made in the 1780s and 1790s to extend mandatory post-execution punishment to other capital crimes, including burglary, highway robbery and some other crimes.[14]

Both dissection and hanging in chains were part of the customary repertoire of sentences that a judge might specify for serious crimes, but their use had been, before the Murder Act, discretionary. There was no legislation or even guidelines about the appropriate use of post-mortem punishment. Post-mortem punishment seems to have been considered by the legislative and judicial Establishment as both a deterrent and an expression of social sanction, even of collective retribution. Peter King has suggested that simple vengefulness might also have played a larger part than is sometimes assumed.

The Murder Act specified that

> [W]hereas the horrid Crime of Murder has of late been more frequently perpetrated than formerly... And whereas it is thereby become necessary that some further Terror and peculiar Mark of Infamy be added to the Punishment of Death, now by Law inflicted on such as shall be guilty of the said heinous Offence;... Sentence shall be pronounced in open Court, immediately after the Conviction of such Murderer... in which Sentence shall be expressed, not only the usual Judgment of Death, but also the Time appointed for the Execution thereof, and the Marks of Infamy hereby directed for such Offenders, in order to impress a just Horror in the Mind of such Offender, and on the Minds of such as shall be present, of the heinous Crime of Murder.
>
> And after Sentence is pronounced, it shall be in the Power of any such Judge, or Justice, to appoint the Body of any such Criminal to be hung in Chains; but that in no Case whatsoever, the Body of any Murderer shall be suffered to be buried, unless after such Body shall have been dissected and anatomized.[15]

In practice, this usually meant that a judge sentencing a murderer would specify that, following execution, the criminal's body be sent to the

---

[14] Richard Ward (2014) 'The Criminal Corpse, Anatomists and the Criminal Law: Parliamentary Attempts to Extend the Dissection of Offenders in Late Eighteenth-Century England', *Journal of British Studies*, 53: 4.

[15] 25 Geo II c. 37. An Act for Better Preventing the Horrid Crime of Murder.

appointed surgeon or anatomist for dissection, or hung in chains. The wording of the Murder Act itself is a little unclear about whether the sentence had to be anatomisation, with the proviso that such a sentence could later be modified to hanging in chains, or whether the judge was empowered at the point of sentencing to specify hanging in chains. At a meeting held on 7 May 1752 for the purpose of resolving any ambiguity, a number of judges argued that hanging in chains should be specified if no surgeon could be found to dissect the body.[16] An initial sentence of dissection was sometimes later changed to hanging in chains at the end of the Session in which the case was tried.

So it was under this legislation that Tom Otter's shocking crime was dealt with. Although the majority of those condemned under the Murder Act in the period between the Murder Act and the Anatomy Act were sentenced to dissection, in a minority of cases the judge specified that the felon be gibbeted, or as it was generally described at the time "hung in chains". Of the 1150 convictions under the Murder Act in England and Wales between 1752 and 1832, 908 (79%) were anatomised and dissected after execution, and 147 (13%) hung in chains. Of the rest, 93 (8%) were pardoned, and two died in prison before the sentence was carried out (Table 1.1 and Appendix y).

OTHER POST-MORTEM PUNISHMENTS: FROM CUSTOMARY SANCTION TO THE FULL FORCE OF THE LAW

Dissection and gibbeting were not the only ways in which social sanction was physically expressed through actions on the dead body. Without any recourse to law, there were mechanisms within the local moral economy by which the status of the deceased could be signalled and reproduced. The purity of unmarried girls, and sometimes boys too, was acknowledged by burying them with a "maiden's crant" or decorative crown.[17] The location of the grave was also to some extent indexical of social standing. Disapprobation could be expressed through denial of a

---

[16] Judges' resolution on the Manner of Sentencing under the Murder Act—National Army Museum Archives, ref. 6510–146(2), 7 May 1752.

[17] Rosie Morris (2013) 'Maiden's garlands: a funeral custom of post-Reformation England', in C. King and D. Sayer (eds.) *The archaeology of post-medieval religion* (Woodbridge: Boydell).

Table 1.1 Numbers hung in chains under the Murder Act

Period	Hung in chains under the Murder Act	Hung in chains for other crimes	Total hung in chains	Hung in chains in each period as percentage of total, 1752–1826 (%)
1752–1776	62	28	90	41
1777–1801	67	48	115	53
1802–1826	12	2	14	6
Total	141	78	219	100

grave space in the desirable areas of the churchyard. The unfashionable north side of the churchyard was the customary burial place of non-communicants, unbaptised babies, strangers and criminals. In some parts of Britain, special burial grounds were kept for the disposal of unbaptised children, foreigners, suicides and criminals, although this practice was not widespread outside Ireland and the northwest of Scotland.[18] Though never formalised in law, burial outside the churchyard or in less prestigious parts of the churchyard was part of the moral economy of the community until the twentieth century.

There were, however, four other kinds of prosecution beside murder that could result in some form of post-mortem punishment: high treason; petty treason; piracy and other crimes on the high seas (these were tried by the Admiralty courts); and the most serious property offences, principally highway robbery and robbery of the mail. Post-mortem treatments of those executed for major property crime, when that sentence was passed, were similar to post-mortem treatments of those executed for murder. Capital criminals convicted by the Admiralty courts also faced punishments similar to those convicted of murder, with the notable feature that they were more likely to be gibbeted and that Admiralty gibbetings had some differences in practice to those convicted in assize courts. High and petty treason, however, were punishable during the

---

[18] E. Murphy (2008) 'Parenting, child loss and the cilline of post-medieval Ireland', in M Lally (ed.) *(Re)Thinking the little ancestor: new perspectives on the archaeology of infancy and childhood* (Oxford: Archaeopress); S. Tarlow (2011) *Ritual, belief and the dead in early modern Britain and Ireland* (Cambridge: Cambridge University Press), pp. 45–52; M. McCabe (2010) 'Through the backdoor to salvation: infant burial grounds in the early modern Gaidhealtachd'. Paper presented at the 32nd Annual Conference of the Theoretical Archaeology Group, University of Bristol, 17–19 December 2010.

long eighteenth century by various kinds of aggravated execution which involved subjecting the body to additional elements of pain and indignity both during and after execution.[19] These post-mortem punishments might more aptly be considered aggravated executions and indeed as the period progressed, some elements of punishment which had previously been carried out on the living body as part of the process of execution were later visited on the newly dead body instead. In addition to these, the crime of suicide—which could not be prosecuted or tried for obvious reasons—was frequently punished by visiting extra humiliations on the dead body.

Crimes Other Than Murder: Treason

Those convicted of treasonable offences were customarily subject to particularly excruciating and slow forms of death. It is widely believed that in Britain treason is still punishable by death. In fact, the death penalty even for treason was abolished in 1998, and no person has been executed for treason in this country since 1946. However, capital punishment remained, in theory, mandatory for high treason even after the death penalty had been abolished for most other offences, evidencing the particular gravity of treason in British law.

Treason offences were divided into high treason, which is treachery against the State or monarch, and petty treason: treachery of a subordinate against their natural or social superior, which would include the murder of an employer by their servant, for example, or of a husband by his wife. It was decided soon after the Murder Act that petty treason came within the purview of the Murder Act, although until the Treason Act of 1791 the traditional means of execution for women convicted of that offence—burning—was used as late as 1788.[20] However, traitors were also subject to special treatments of the body.

Well into the nineteenth century, the official legal punishment for male traitors was to be "hung, drawn and quartered", which involved removing the traitor's body from the scaffold before he was dead and cutting out his entrails before his own eyes. Finally, he was beheaded

---

[19] Peter King (forthcoming) *Punishing the Criminal Corpse*.

[20] Margaret Sullivan was burned for petty treason in 1788. Gatrell *The Hanging Tree*, pp. 337–38.

and his body divided into quarters, which could be displayed in a public place. For women, including those found guilty of petty treason, the legal execution for treason was by burning at the stake. However, by the eighteenth century, it had become normal practice to kill traitors first by hanging (for men) or strangling (for women), so that then being burned or disembowelled became a post-execution punishment.[21]

The traditional fate of the traitor's body was for his quarters to be disposed "At the King's pleasure". Until the eighteenth century, this generally meant displaying the heads of traitors at city gates or on prominent public buildings. Other body parts, being less recognisable, were less frequently displayed.

During the period of the Murder Act, the display of traitors' heads and quarters was definitely less common in Britain than it had been in the early modern period, and the times and places where it was in more frequent use—Ireland through much of the eighteenth and nineteenth centuries and Scotland in the wake of the 1745 rebellion—were those where the sovereignty of the monarch and the rule of Parliament were most seriously threatened.[22] Following the Jacobite rebellion, there were 79 executions for treason in 1746, in London, York, Carlisle, Brampton and Penrith. Although as traitors their bodies could be decapitated, quartered and displayed, letters at the time show that at least some of those executed in Cumberland were immediately buried.[23] However, 18 of those considered most culpable were brought to London for trial and execution, and their fates are better recorded. Their bodies were hanged, drawn and quartered and then beheaded. Although the bodies appear to have been buried afterwards, at least some of the heads were retained and displayed. Francis Towneley's body, for example, was buried in St Pancras churchyard, but his head was placed on a spike at Temple Bar, next to that of fellow Jacobites George Fletcher and Thomas David Morgan. The head of Thomas Deacon, who was executed the same day, was pickled and

---

[21] Beattie *Crime and the Courts*, p. 451.

[22] J. Kelly (2015) 'Punishing the dead: execution and the executed body in eighteenth-century Ireland', in R. Ward (ed.) *A Global Gistory of Execution and the Criminal Corpse* (Basingstoke: Palgrave); Rachel Bennett (2015) *Capital Punishment and the Criminal Corpse in Scotland 1740 to 1834*, Unpublished Ph.D., University of Leicester.

[23] Bennett, *Capital Punishment and the Criminal Corpse in Scotland*.

transported to Manchester and Carlisle to be exhibited. Exhibited heads were sometimes rescued: Towneley's head was recovered from Temple Bar and interred in the family vault at Towneley Hall in Burnley.

In practice, after the executions of the Jacobite rebels of 1745, there were only two instances of disembowelling as a formal punishment for treason—those of Francis Henry La Motte in 1781 and David Tyrie in 1782. Although the sentence pronounced continued to condemn the prisoner to be "hanged by the neck but not until you are dead, but that you be taken down again, and that while you are yet alive, your bowels be taken out and burnt before your faces, and that your bodies be divided each into four quarters, and your heads and quarters be at the King's disposal", in practice the executioner had discretion to waive the disembowelling and quartering and to abbreviate other elements. Even La Motte had hanged for nearly an hour before he was disembowelled, so he would have been deeply unconscious, if not dead, by the time that part of his sentence was carried out. Thus, by the late eighteenth century, burning, disembowelling and so on had become effectively post-execution punishments.

Executed in Hampshire in 1782, David Tyrie might have been the last person to be given the full works. Tyrie was convicted of carrying on a treasonous correspondence with the French and had some association with De La Motte, executed the previous year. *The Hampshire Chronicle* reported on 31 August of that year, "His head was severed from his body, his heart taken out and burnt, his privities cut off, and his body quartered. He was then put into a coffin, and buried among the pebbles by the seaside; but no sooner had the officers retired, but the sailors dug up the coffin, took out the body, and cut it in a thousand pieces, every one carrying away a piece of his body to shew their messmates on board". Interestingly, although Tyrie was given the whole medieval gory horror, his head and quarters were not piked and displayed but buried on the shore, a treatment normally accorded to suicides and strangers. De la Motte's treatment was slightly more lenient: his body was only symbolically scored rather than fully quartered. His body was placed immediately in a coffin by an undertaker, but the head was "reserved by the executioner to be publicly exposed".[24]

---

[24] J. Williams (1781) *The life and trial of F.H. de la Motte, a French spy, for high treason* (London: T. Truman), p. 34. *The Newgate Calendar*, however, says that the head was placed with the body in the coffin.

James O'Coigley, executed in Kent in 1798 for high treason, was beheaded after death, although this was carried out by a surgeon rather than the executioner. Both head and body were immediately put into a coffin and buried.

The old sentences were enacted only a few times in the nineteenth century. The Despard conspirators were decapitated in 1803, though not disembowelled or quartered, and their heads do not seem to have been retained for display after being shown to the crowd.[25] In 1812, two men—John Smith and William Cundell—were hanged and beheaded for treason, following their desertion from the British to the French army. Their heads were shown to the crowd but then returned with their bodies to their friends for burial.[26] The leaders of the Pentrich revolt were executed in 1817. They were sentenced to be hanged drawn and quartered, although in the event quartering was waived. After they were dead, they were beheaded and then "buried in one grave in St Werburgh's churchyard".[27] Finally, in 1820, the Cato Street conspirators were hanged and then beheaded[28] by a surgeon. Three other would-be Scottish rebels were executed at Glasgow and Stirling later the same year; there were no further judicial beheadings in Britain.

The bodies and heads—of the Cato Street conspirators were not exhibited, nor were they returned to the men's families, who had petitioned to be allowed to claim them. Instead, they were buried within the prison compound, covered in quicklime. The wives' petitions were not purely sentimental or dutiful; according to Gatrell, they proposed to exhibit the bodies commercially to raise money for the conspirators' families.[29] By the time of the Cato Street executions, therefore, the exhibition

---

[25] C. Oman (1922) 'The Unfortunate Colonel Despard' in *The Unfortunate Colonel Despard and other studies* (London: E. Arnold), pp. 21–22.

[26] *The Criminal recorder: or, Biographical sketches of notorious public characters, including murderers, traitors, pirates, mutineers, incendiaries ... and other noted persons who have suffered the sentence of the law for criminal offenses; embracing a variety of curious and singular cases, anecdotes, &c*, Vol. 2 (London: J. Cundee, 1815), pp. 288–96.

[27] P. Taylor (1989) *May the Lord have mercy on your soul: murder and serious crime in Derbyshire 1732–1882* (Derby: JH Hall and sons), pp. 37–39.

[28] The execution of the Despard conspirators and the Cato Street conspirators is extensively described and discussed by Gatrell in *The Hanging Tree*, pp. 298–321.

[29] Gatrell *The Hanging Tree*, p. 308.

of the heads or the bodies of traitors was not carried out, either for private profit or for public statement.

Interestingly, the only criminals to stand trial posthumously in the post-medieval period were charged with treason. In England, Oliver Cromwell, Henry Ireton and John Bradshaw were tried posthumously for treason in 1661 and, on being found guilty, were exhumed and punished by hanging, beheading and the display of their heads. The remains of Robert Leslie, accused of treason in the Scottish courts in 1540, were allegedly exhumed before the trial, and his bones were brought to the dock, but no similar case happened in England.[30]

The punishment of traitors' bodies can be mostly fitted to a broad tripartite chronological division: first is the medieval and early modern tradition of aggravated execution with extreme pain and, essentially, torture. This was part of a broad European tradition of spectacular pain, famously exemplified in Foucault's description of the death of Damiens the regicide in 1757.[31] This was succeeded in the eighteenth century by a period during which execution by, effectively, public torture gave way to a public execution which reserved the spectacular elements of burning, dismemberment and public display to the treatment of the post-mortem body.[32] Indignity and disintegration of the body (psychological and social distress) thus supplanted pain (physical distress) as the most severe punishment. Finally, over the course of the later eighteenth and nineteenth centuries, public humiliation of the body was succeeded by private and increasingly efficient, physical punishment. The disposition of quarters and display of heads ended, and the practices of gibbeting,

---

[30] The case of Robert Leslie was cited in the *Encyclopedia Britannica* of 1904 and is repeated in a number of twentieth-century sources without attribution. Court records of December 1540 seem to suggest only that Leslie's wife and children were summoned to appear in his stead. S. Tarlow (2013) 'Cromwell and Plunkett: two early modern heads called Oliver', in J. Kelly and M. Lyones (eds.) *Death and dying in Ireland, Britain and Europe: historical perspectives* (Dublin: Irish Academic Press), pp. 59–76.

[31] Michel Foucault (1991) [orig. Paris: Gallimard, 1975] *Discipline and Punish* (London: Penguin).

[32] A further twist is that the body removed from the gallows following a strangulation hanging was often still alive though unconscious. The frequency with which hanged 'dead' bodies revived on the dissection table testifies to the inexactitude of pre–long-drop hanging. See E. Hurren (2013) 'The dangerous dead: dissecting the criminal corpse' *The Lancet*, 27 July 2013, Vol. 382, pp. 302–03.

public dissection and eventually public execution of any kind were gradually abandoned between the late eighteenth and mid-nineteenth centuries. Even traitors were thenceforward executed privately by the quick and efficient long-drop method, and their bodies buried within prison walls.

This kind of chronology of punishment is observed in not only the case of treasonous bodies but also other kinds of criminal. The changes are to do with cultural attitudes as well as the law.

That the disembowelling and beheading of traitors feels anachronistic in the eighteenth and nineteenth centuries is not a new point. It is both in the spectacular pain of prolonged, multi-stage executions and in the superfluity of post-mortem shaming of the body that the traitor's death claims a medieval descent. Yet the extensive, irrational, spectacular punishment of the body was also the core of the post-mortem punishments of the 1752 Murder Act. King's review of the published debate about aggravated forms of capital and corporal punishment demonstrates that, although executions and publically bloody punishments declined in number during the eighteenth century, they actually increased in brutality up until the 1770s. For King, the Murder Act is not an aberration but the culmination of a series of debates. This presents a different kind of eighteenth century, one that is very different from Norbert Elias's civilising journey, and challenges progressivist histories that emphasise the spread of humane and empathetic attitudes.[33]

Crimes Other Than Murder: Suicide

Post-mortem treatment of the body could be used as a means of expressing social sanction for a range of deviant behaviours, including criminality, even without being formalised in law. This is most notable in the treatment of suicide bodies. The practice of giving special burial treatment to suicides was well established in Britain since at least the medieval period. In early modernity, under the influence of puritanical and fundamentalist Protestantism, suicide was considered to be evidence of the sin

---

[33] Norbert Elias (1994) *The civilising process*. Oxford: Blackwell. Elias offers a long-term history of manners by which self-restraint, circumspection and 'civility' came to characterise social and political relationships over the second millennium AD.

of despair and almost invariably thought to be the result of succumbing to diabolical temptation. By the end of the eighteenth century, however, ordinary people throughout Europe were far more likely to want to see suicide as the result of mental illness and to try to circumvent traditional, religious or legal requirements that suicides be denied normal burial.[34] However, attitudes towards taking one's own life show considerable variation even in the eighteenth century and were affected by the circumstances of the suicide.

Throughout the eighteenth and nineteenth centuries, suicide was considered a crime under both secular and canon law. Those who committed suicide in order to escape the justice of the State were double criminals. Since the means of death had been taken from the State, other forms of punishment were placed upon the suicide, foremost among which were post-mortem punishment of the body and forfeiture of the Estate. As Houston notes, forfeiture was "a token of blame and of 'apology'", but the punishment of the body was both more shameful and more punitive.[35] MacDonald and Murphy's history of suicide records that the normal punishment for suicides until 1823 was forfeiture and profane burial. The 1823 Act ended the custom of profane burial for suicides, but it is noteworthy that profane burial was never a universal and legally enshrined rule: the 1823 act only put a stop to a local customary practice which had already fallen out of use in many parts of the country, as a more sympathetic attitude to suicides gained ground. In fact, Houston contends that profane burial in the form of highway burial with a stake through the body was predominantly a southeast English custom and that widely variable practices are described in provincial newspaper and legal accounts of the disposal of the suicide's body. Houston notes, for example, that in 50 years of the *Cumberland Pacquet* only 3 of 18 suicides reported in the northern counties of England were linked to unusual burials: one staked at a crossroads, one on Lancaster Moor and one buried at Low Water mark. All three are from 1790–1791 and might

---

[34] MacDonald and Murphy, *Sleepless Souls*. See also the essays in Jeffrey Watt (ed.) (2004) *From Sin to Insanity* (Ithaca: Cornell).

[35] The history of suicide in Britain in the eighteenth and nineteenth centuries has been most comprehensively addressed by MacDonald and Murphy *Sleepless Souls* (1999) and Rab Houston *Punishing the dead* (2010). The literature on the legal, theological and social context of suicide in history is vast and complex; here we concentrate only on the fate of the body.

reflect a particular moment of public anxiety about self-murder. Two more staked burials of suicides from other counties were mentioned in the *Paquet*, and a few more mention unusual locations, but of a total of 209 reported suicides nothing is mentioned of the disposal of the body in the majority of cases.[36]

The prevalence of staked highway burial is hard to estimate. Historical sources have not been systematically reviewed for much of the country and are in any case not always informative. Even where a coroner's court recommended staked highway burial, actual practice is not often attested: to our knowledge, there is no coroner's court equivalent of the sheriffs' cravings that detail actual expenditure. Archaeological evidence is an excellent source but very few suicide burials are known. In particular, highway burials, by virtue of their very exclusion from normal burial places, are not generally anticipated when road development schemes are carried out, and it is likely that many or most have been destroyed in twentieth-century road construction programmes without any kind of archaeological excavation or recording having taken place. The skeletons of bodies buried without coffins rarely survive for two hundred years except as fragments and stains,[37] and if such remains were excavated without archaeological training or using archaeological methods, they would be very unlikely to be noted or recorded. Halliday's short article on criminal graves has little sense of chronology and does not distinguish suicides from other executed criminals.[38] It is interesting, however, that nearly all the cases of crossroads burial he mentions are from the south and east of England. The one Welsh case discussed—reported in the *Gentleman's Magazine* in 1784—was buried on the shore, disregarding the coroner's suggestion that she be given staked crossroads burial.

The desecration of suicides' bodies and the enactment of practices designed to appease the spirit or lay the ghost of a suicide were not ordered or sanctioned by the Church of England, although religious authorities did insist from time to time that suicides not be given full and normal burial rites.[39] Nor, as we have seen, did English law insist on their special treatment.

---

[36] Rab Houston, *Punishing the Dead*, p. 203.

[37] Sian Anthony (2015) 'Hiding the body: ordering space and allowing manipulation of body parts within modern cemeteries', in S. Tarlow (ed.) *The archaeology of death in post-medieval Europe* (Berlin: DeGryuter Open), pp. 172–90.

[38] Halliday, 'Criminal graves and rural crossroads'.

[39] MacDonald and Murphy *Sleepless souls*, pp. 42–43.

Houston's contention is that suicide burial customs were regionally and chronologically variable and indeed were not necessarily standard even within a small area. So the degree of "profanity" in a profane burial might be quite varied. Since practice was not specified authoritatively by Church or State, suicide burial might serve a number of purposes. Briefly, these could include the following:

1. Punitive practice as part of the retributive process. To express social sanction
2. Deterrence. In Weever's often-cited words "to terrifie all passengers, by that so infamous and reproachfull a buriall, not to make such their finall passage out of this world"[40]
3. Preventing the ghost of the suicide from returning to trouble the living, through pinning (with a staked burial) or burial at a crossroads (which, it has been suggested, would confuse and disorientate the revenant)
4. Exclusion from the community of the dead. This was enacted spiritually in the exclusion of suicides form normal rites and normative daytime burials and spatially in keeping the place of suicide burial separate from the normative cemetery. They were buried either outside the churchyard or on its inauspicious north side.

Until the decriminalisation of suicide in 1961, all suicides except those who were insane were criminals.[41] But some suicides were criminals twice over. Those men and women who evaded the noose, gaol, transport or other public retribution by taking their own lives were a special—and, it was often opined, particularly culpable—kind of suicide. The most famous criminal suicide of our period was the death of John Williams in Coldbath Prison, London, in 1811, while he was awaiting trial for the Ratcliffe Highway murders (although some doubt has been raised about whether Williams's death was indeed a suicide).[42]

---

[40] John Weever (1631), *Ancient and Funerall Monuments with in the united Monarchie of Great Britaine, Ireland and the Islands adjacent* (London: Thomas Harper), p. 22.

[41] Suicide Act 1961 (9 & 10 Eliz 2 c 60).

[42] Thanks to Steve Poole for drawing my attention to the possibility that Williams did not take his own life.

John Williams's burial was pure pageant. His body was taken from the prison where he died, laid out on a board next to the blood-stained tools with which he had murdered his victims. The board was put into a cart and followed by a crowd of up to 20,000 people through the streets of London. The route taken by the wagon passed the houses of his victims, at each of which the procession halted. Eventually, the procession reached the Cannon Street crossroads, where the body was stuffed into a grave that was slightly too small and a stake driven through it.[43]

THINKING ABOUT GIBBETS: THE HISTORIOGRAPHY OF HANGING IN CHAINS

"On the edge of the river I could faintly make out the only two black things in all the prospect that seemed to be standing upright; one of these was the beacon by which the sailors steered—like an unhooped cask upon a pole—an ugly thing when you were near it; the other, a gibbet with some chains hanging to it which had once held a pirate".[44]

Hanging in chains, then, was only one way among several of expressing social or judicial censure after death, and it occurred more rarely than staked burial or dissection. However, gibbetings left a cultural mark in the minds and landscapes of those who witnessed one, that was perhaps disproportionate to their frequency.

Given the emotional impact of the real or imagined presence of the gibbet (young Pip's awareness of the pirate's gibbet on the marsh in the first chapter of *Great Expectations*, for example), there is surprisingly little sustained or academic study of the practice. This contrasts with the large body of literature on dissection as a post-mortem punishment.[45] The two most extensive and detailed studies of the practice, William Andrews *Bygone Punishments* (1899) and especially Albert Hartshorne's *Hanging in Chains* (1893), are both more than a hundred years old,

---

[43] Newgate Calendar (http://www.exclassics.com/newgate/ngintro.htm).

[44] Charles Dickens (1996 [1860–61]), *Great Expectations* (London: Penguin), p. 7.

[45] See Ruth Richardson (1989) *Death, Dissection and the Destitute* (London: Routledge & Kegan Paul); Elizabeth Hurren (2012) *Dying for Victorian Medicine: English Anatomy and its Trade in the Dead Poor, c. 1834–1929* (Basingstoke: Palgrave Macmillan); Thomas Laqueur (1989) 'Crowds, Carnival, and the State in English Executions, 1604–1868', in Lee Beier, David Cannadine, and James Rosenheim (eds.) *The First Modern Society: essays in honour of Lawrence Stone* (Cambridge: Cambridge University Press).

and neither makes any attempt to be exhaustive or systematic or to put the practice into much historical context.[46] Hanging in chains is often mentioned by crime historians as a sentence, but the technicalities of the physical process, the criteria by which gibbets were located in the landscape, and the material impact of their presence have not been subject to analysis, nor have the contrasts between gibbeting and dissection been discussed or explained. This book attempts to draw out the main features of gibbeting, principally during the period of the Murder Act. This chapter reviews the broad historical context of gibbeting under the Murder Act: how frequent was the practice and how did it change over time? What kinds of crime or criminal were most likely to be punished in that way? It also corrects some widespread misunderstandings about hanging in chains. The second chapter is concerned with questions of geography and the events of a gibbeting itself: where were gibbets sited? Which parts of the country were keenest on the practice? How were the precise locations of gibbets determined? What actually happened when a person was hung in chains? What were the technical and material features of the apparatus? The third chapter takes us beyond the original occasion of the gibbeting to look at the afterlives of gibbets—how did they shape the landscape and people's experience long-term? When and why were they taken down and what happened to the remains and the material then? The book ends with some consideration of why this punishment, which seems in some ways anachronistically brutal in the later eighteenth century and certainly was more costly than its alternative (dissection), continued to be carried out.

Who Was Hung in Chains?

Although the Murder Act dealt specifically with murder, gibbeting and dissection were sometimes specified for other crimes too. Next to murderers, the most likely to be hung in chains were those who came before the Admiralty courts (mostly for killing offences, piracy or smuggling), highway robbers and those convicted of robbing the mail (Table 1.2). The practice of hanging highway robbers in chains near the scene of

---

[46]W. Andrews (1899) *Bygone Punishments* (London: William Andrews and Co); Albert Hartshorne (1893), *Hanging in Chains* (Cassell, New York).

Table 1.2 Crimes punished by hanging in chains, 1752–1832

Hanging in chains for all categories of offence, 1752–1832		
Offence	Number	Percentage (%)
Murder (including Admiralty cases)	144	64.9
Mail robbery	31	14.0
Admiralty offences (not including murder)	23	10.4
Highway robbery	10	4.5
Burglary and housebreaking	7	3.2
Robbery	2	0.9
Shooting with intent to kill	2	0.9
Animal theft	1	0.5
Arson	1	0.5
Riot	1	0.5
Total	222	100.0

their crime was apparently well established by the time of the Murder Act. As early as 1694, a proposal to formalise the practice had been put to Parliament, and Cockburn has found evidence that by 1770 it was normal for a Post Office official to attend the trial of a mail robber to remind the judge that hanging in chains was the customary sentence in such cases, or to pressure the Secretary of State to order that punishment if the judge was not willing to be guided.[47] Harper says that as a result of intervention by the Earl of Leicester, Postmaster General at the time, after 1753 those found guilty of robbing the mail were to be gibbeted after execution.[48] However, despite the existence of a few personal letters requesting a sentence of gibbeting in individual cases, there is no universal legislation or general guideline extant. There are, however, records of the Postmaster General applying on specific occasions for the body of a mail robber to be hung in chains. For example, Lord Sandwich requested in April 1770 that the body of John Franklin, convicted of the robbery of the Bristol mail, be hung in chains. The judge turned down his request on the grounds that the robbery had not involved violence, but Sandwich went over his head to the High Sheriff to procure an order that Franklin's body be hung in chains near the place where the robbery was

---

[47] J.S. Cockburn (1994) 'Punishment and Brutalization in the English Enlightenment' *Law and History Review* 12(1): 155–79, p. 167.

[48] G. Harper (1908) *Half-hours with the Highwaymen; picturesque biographies and traditions of the knights of the road (Vol. 1)* (London: Chapman and Hall), p. 206.

committed. Interestingly, in this case, the Postmaster General offers no other reason for his request than that gibbeting "had always been done in cases of mail robberies".[49] It was thus perceived traditional practice rather than any motivation articulated in a legal act that perpetuated the custom of gibbeting mail robbers near the scene of their crime. The most frequent crimes other than murder for which gibbeting was a punishment were all capital crimes which threatened the orderly administration of the capitalist state (although forgery does not seem to have been punished in this way unless the criminal was also found guilty of other serious crimes). It could thus be suggested that crimes against the State were more likely to lead to the spectacular punishment of hanging in chains than private, personal or domestic, but equally serious, crimes against the person or burglary, which might be more likely to receive a sentence of dissection.

Smugglers

In the period immediately preceding the Murder Act, a large number of men were hung in chains for smuggling. Between 1747 and 1752, 50 people were convicted of smuggling in the counties of Sussex and Kent, of whom 42 were hanged, and 16 of those were also hung in chains. There was clearly regional variation at play here also since none of the 23 smugglers convicted in East Anglia over the same period was sentenced to any post-mortem punishment at all.[50]

INTERPRETING THE MURDER ACT: DISSECTION OR HANGING IN CHAINS?

Whether a convicted murderer should be dissected or gibbeted was left to the discretion of the judge, as was the inclusion of post-mortem punishment in the sentence of those found guilty of other crimes.

The rationale for deciding which people should be dissected and which hung in chains is much harder to understand. When Thomas Hanks was hung in chains in Gloucestershire in 1763 instead of being

---

[49] State Papers, Southern Department SP 44/89/350.

[50] Zoe Dyndor (2015) 'The Gibbet in the Landscape: locating the criminal corpse in mid-eighteenth-century England', in R. Ward (ed.) *A Global History of Execution and the Criminal Corpse* (Basingstoke: Palgrave).

dissected as originally specified, the local newspaper reported only that such a punishment would be "better".[51] At the Hereford Lent Assizes in 1770, all of the six men found guilty of the murder of William Powell and sentenced to death were destined by the judge for dissection,[52] but ultimately only four were dissected: William Spiggott and William Walter Evan were hung in chains instead.[53] Pamphlet accounts of their crime and trial give no reason for this differential treatment—and the two men gibbeted were neither more nor less culpable than those dissected. A similar situation arose following the conviction of three men— John Croxford, Benjamin Deacon and Richard Butlin—for murder at the Northamptonshire assizes on 31 July 1764. Although the original sentence was that all three should be sent for dissection under the terms of the Murder Act, a warrant from the judge to the sheriff records a subsequent decision that Croxford alone should be hung in chains instead.[54] Indeed, of 16 people sentenced to be dissected in Northamptonshire between 1739 and 1832, at least five were ultimately hung in chains instead. Edward Corbett, convicted of murder at the Buckinghamshire Assizes in 1773, was sentenced to be dissected, but his sentence was amended to hanging in chains because, according to the Assize Calendar, "no surgeon is willing to receive the said body". Similarly, when William Suffolk was executed in Norfolk in 1797, no surgeon came forward to claim the body, so the court ordered instead that it be hung in chains "near as may be where the said felony was perpetrated"[55]; and Thomas Otley, executed for murder in 1752 in Suffolk, was "ordered to be hanged in chains (no surgeon be willing to receive his body) pursuant to the statute in such case lately made".[56] In Suffolk in 1783, James May and Jeremiah Theobald were both convicted of murder and sentenced to hanging and dissection. However, both bodies were instead hung in chains at Eriswell, the scene of crime "at the request of the prosecutor", according to a pamphlet detailing their trial, although no further

---

[51] N. Darby (2011) *Olde Cotswold Punishments* (Stroud: History Press), p. 24.
[52] *General Evening Post*, 31 March–3 April 1770, issue 5690.
[53] *Independent Chronicle*, 11–13 April 1770, issue 85.
[54] TNA E389/243/410.
[55] TNA E389/250/79 (Assize Calendar Norfolk 21 March 1797).
[56] Sheriffs' Cravings Suffolk 1752.

explanation of this decision is given.[57] The same happened nine years later at the same assize court in the case of Roger Benstead,[58] again with no reason given, although a contemporary account notes that this part of the sentence seemed to affect the condemned with a greater dread than any other aspect of the sentence, including the execution itself.[59] In 1794, John and Nathan Nichols, father and son, were both found guilty of the same murder, also in Suffolk, and originally both sentenced to be sent to the surgeons.[60] However, after execution, the older man's body was hung in chains whereas the younger man was dissected.[61]

In researching this book, we were for some time puzzled by the frequency with which the judge appeared to have changed his mind about what kind of post-mortem provision should be applied. We encountered numerous cases where before the judge left town he directed that an offender should be hung in chains rather than dissected. Such voltes-faces never occurred the other way round (from hanging in chains to dissection). The initially mystifying practice of substituting the gibbet for the scalpel at what appeared to be the last minute was explained by another piece of documentary evidence. The discovery of a recorded meeting of all circuit justices shortly after the passage of the Murder Act shows this practice to be an interpretation of the consensus reached there that the proper sentence was normally to be hanging until dead followed by delivery to a surgeon for dissection and anatomisation. The order to hang in chains was to be made as an amendment to the sentence delivered in open court.[62] On many occasions, this seems to have occurred as part of the "dead letter"—the instructions left by the judge at the end of an assizes listing which sentences

---

[57] *The Trial at Lage of Jeremiah Theobald, otherwise Hassell, and James May, otherwise Folkes* (Ipswich: Shave and Jackson) 1783.

[58] Richard Deeks (1984) *Some Suffolk Murders* (Long Melford: R&K Tyrell), pp. 10–11.

[59] *The trial of Roger Benstead the elder* (Bury St Edmunds: P. Gedge) 1792, p. 14.

[60] *The trial of John and Nathan Nichols, (Father and Son)* (Bury St Edmunds: P. Gedge), p. 8.

[61] Diary of William Goodwin, surgeon, of Earl Soham Suffolk. Suff RO HD 365/3 vol. 2, from 1791.

[62] Judges' resolution on the Manner of Sentencing under the Murder Act—National Army Museum Archives, ref 6510-146(2), dated 7 May 1752.

of execution were to be actually enacted and who was to be reprieved. The letter would be informed by representations made to the judge based on local knowledge of the accused or attitudes towards their crime. Decisions in the dead letter were not usually explained.

In most cases, then, no reason for hanging in chains rather than gibbeting is given. Where a reason is stated, it relates to those cases where hanging in chains was a pragmatic response to the absence of any surgeon willing to take the body for dissection. Whereas some kinds of body were in high demand for dissection—young and fit ones, large ones, female ones and unusual ones—old, small, white, male ones were less valuable. This may be the reason that no woman was ever hung in chains under the Murder Act—since women were much less likely to be accused of or condemned for murder, female bodies were only rarely available to medical science under the terms of the Murder Act. The bodies of executed women whose crimes fell under the Act were therefore highly prized for dissection. When John Swan and Elizabeth Jeffryes were both convicted of the same murder in 1752, only Swan was hung in chains, but Jeffryes's fate is unclear[63]; and whereas William Winter was hung in chains near Elsdon in Northumberland, the two women convicted alongside him, Jane and Eleanor Clark, were both dissected. Similarly, the decision to hang John Nicholls in chains and dissect his son Nathan might also indicate that the younger, fitter body was of greater interest to surgeons than the body of an old man. In 1759, Surrey surgeons rejected the body of Robert Saxby altogether because he was too old; he was therefore hung in chains instead.[64] Medical interest might also have influenced the post-mortem fate of John Pycraft, who was executed for murder in Norfolk in 1819. Pycraft was affected by some kind of dwarfism. His measurements are given in the *Bury and Norwich Post* of 25 August 1819 as 4'2" in height, with legs of 18", arms of 13.5" and his skull circumference as 23.5". His body was sent for dissection

---

[63] The trial of Swan and Jeffryes took place just before the Murder Act came into force. They were both found guilty—Swan of petty treason and Jeffryes of murder—and both hanged, but it seems that Jeffryes's post-mortem fate was neither the gibbet nor the scalpel. *The Authentick Memoirs of the Wicked Life and Transactions of Elizabeth Jeffryes* (2nd edn., London, 1752) claims that her body was taken away by her friends, as does the *London Evening-Post*, 28–31 March 1752.

[64] *Whitehall Evening Post or London Intelligencer*, 11 August 1759, issue 2091.

and his skeleton retained by the Norfolk and Norwich Hospital museum where it was catalogued under his own name.[65]

Given this context, it is surprising that Toby Gill, "Black Toby", was hung in chains rather than dissected after his conviction for murder in 1750. Convicted for the murder of a local girl, Ann Blakemore, Gill, who was a drummer in Sir Robert Rich's regiment, was gibbeted at Blythburgh in Suffolk. Gill was described at the time as "a black" and would normally therefore have been of interest to the surgeons.

THE RISE AND FALL OF THE GIBBET

For clarity, the term gibbet here is used to describe the whole structure used to display the corpse of a criminal, including post and arm, chains and cage. The framework from which execution by hanging took place is called a scaffold or gallows. During the eighteenth and nineteenth centuries, the terms gibbet and scaffold were sometimes used interchangeably; and "gibbet" could be used loosely to describe the whole edifice, or just the standing post, with the chains and cage described either in those words or together as "irons" or "chains". Variation in the technology and design of the gibbet is discussed in the next chapter. A typical gibbet, however, would comprise a wooden pole of up to twelve metres fixed securely into the ground. It would have a cross arm at the top projecting on one side or sometimes on both sides to make a T shape, usually braced with supporting cross struts. From the end of the arm, a substantial iron hook or socket projected from which was suspended the gibbet cage on a short length of chain. The cage itself was often anthropomorphic and was always made of iron.

The peak popularity of gibbeting in England and Wales was during the mid-eighteenth century, just before the Murder Act in 1752. Figure 1.1 shows the number of gibbetings annually rising to a peak in the 1740s and then declining rapidly. After 1800, there were very few gibbetings in England and Wales; there were no gibbetings at all for property crime after 1803 and very few for murder. Only two people were hung in chains in the 1810s outside the Admiralty courts, and one in the 1820 s. Another man sentenced to be hung in chains in 1827 near Brigg, Lincolnshire, had his sentence remitted following a petition by the

---

[65] NRO NNH 29/2 Catalogue of the Norfolk and Norwich Hospital Museum.

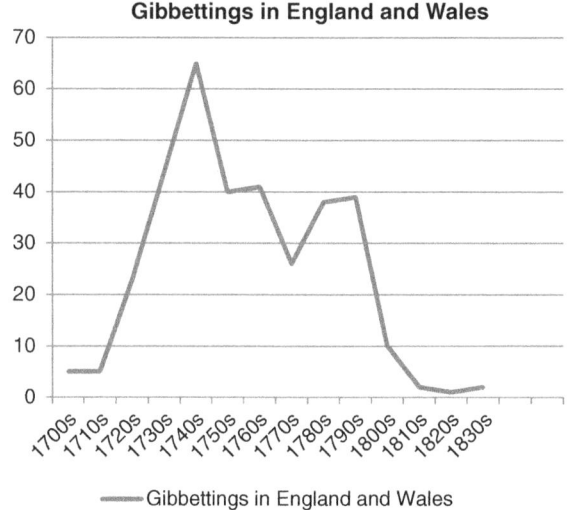

Fig. 1.1 Number of gibbetings per decade in England and Wales, 1700–1832

local inhabitants.[66] The two sentences of gibbeting passed in the summer of 1832, which turned out to be the last hurrah for hanging in chains in Britain, were probably based on a misinterpretation of the Anatomy Act, passed earlier that year, which removed the option of anatomical dissection for convicted murderers. In fact, the more usual alternative of burial within prison grounds was already in use, but it is possible that judges used to passing sentence of dissection believed that gibbeting was the only possible sentence that remained open to them for convicted murderers. There was also a widespread misapprehension at the time that the power to hang in chains had been given to the courts by the Anatomy Act, when the truth was that such powers had never been revoked but had largely fallen into disuse until, in 1832, the Anatomy Act banned what was generally the preferred option. The gibbetings of William Jobling and James Cook that year aroused considerable media interest and a general outcry among the educated classes.

In 1834, the practice of hanging in chains was formally abolished, two years after Parliament ordered that the gibbet of James Cook be taken

---

[66] Andrews *Bygone Punishments*, p. 73.

down from a road junction on the edge of Leicester, only three days after being hung up there. By that stage, there was a very strong feeling that hanging in chains was barbaric and ill-suited to a civilised age. A journalist of the *Leicester and Nottingham Journal* on 18 August 1832 reflected presciently on the dismantling of Cook's gibbet:

> we are glad that the disgusting *sight* has been removed considering it, as we do, the revival of a barbarous custom which a more humanized age has long exploded from the statute book. That the application should have been made in the case of one of the most brutal murders ever committed, is singular; but it will be attended with one important effect. James Cook will be the last murderer that will be sentenced to be hung in chains, since no Judge can hereafter think of awarding the punishment to ordinary murderers while the most atrocious delinquent of that description has been *ungibbeted* by an order bearing the King's sign manual.

It is worth noting, however, that disgust at the sight was not sufficiently widely shared to prevent crowds of more than 20,000 attending Cook's gibbeting.[67]

During the debate accompanying the first presentation of the motion to end gibbeting in 1834, one M.P. pointed out that a judge in Ireland had "only the other day" ordered a murderer to be dissected, despite the official cessation of that form of post-mortem punishment two years earlier, because he considered it "preferable" to hanging in chains.[68] The history of gibbeting in Ireland follows a different trajectory to the English story. Hanging in chains was still widespread in early nineteenth-century Ireland, perhaps because it was valued as an exemplary punishment for crimes with an element of sedition or those judged to threaten the orderly functioning of the State. In England, these include the crimes of piracy, smuggling and mail robbery; in Ireland, crimes which imperilled the tenuous grip of British control were more likely to be punished by spectacular treatments of the body, such as hanging in chains. The landscape of County Louth in Ireland, notable to the British as a breeding ground of sedition and a threat to the authority of the State, was described around 1816 as being "studded with gibbets" containing the remains of Ribbonmen, a group of anti-English Irish Catholics, set

---

[67] *Leicester and Nottingham Journal*, 18 August 1832.
[68] Hansard *HC Deb 13 March 1834, vol. 22, cc155–7*.

up near the homes of those convicted (in Carleton's vivid account, the tarred sacks containing the remains of the executed Ribbonmen attracted so many flies that the sound of buzzing could be heard some distance away).[69] Although the overall capital conviction rate in Ireland was lower than in England, executions which severely damaged the body and caused extensive pain were comparatively more frequent. Bodies were gibbeted in Ireland fairly commonly during the eighteenth century, despite public unease which Kelly attributes both to disgust at the smell and sight of decaying bodies, especially in built-up areas, and to religious and ethical scruples. It may be that ambivalence about the post-mortem exhibition of the body was more pronounced in Catholic countries, although there is no doctrinal reason why this should be the case.[70]

Some Common Misconceptions

The technical and geographical details of gibbeting will be reviewed in the next chapter, but first it is worth correcting or clarifying some widespread misapprehensions about hanging in chains, arising mostly from popular or secondary sources.

Myth 1: Gibbeting Is the Same as Execution by Hanging

While gibbet can be a synonym for gallows or scaffold, gibbeting refers only to the practice of displaying the dead (or, exceptionally, dying) body in a suspended device. In this book, I refer to the structure used for carrying out executions by hanging as the scaffold or gallows and use the term gibbet to refer only to the cage and its pole. Sometimes, particularly in parts of southern England, criminals were executed at the scene of their crime, although this practice had declined in popularity by the time of the Murder Act.[71] When this happened, the criminal would be

---

[69] W. Carleton (1894) *The life of William Carleton being his autobiography and letters; and an account of his life and writings, from the point at which the autobiography breaks off*, edited by David J. O'Donoghue, p. 134.

[70] J. Kelly (2015) 'Punishing the dead: execution and the executed body in eighteenth-century Ireland', in R. Ward (ed.) *A Global History of Execution and the Criminal Corpse* (Basingstoke: Palgrave).

[71] S. Poole (2008) 'A lasting and salutary warning': incendiarism, rural order and England's last scene of crime execution'. *Rural History* 19: 163–77.

hanged from a temporary scaffold and then taken down, encased in a gibbet cage and hoisted back onto the same structure.

Myth 2: Gibbeting Involves Leaving People to Die in an Iron Cage

Popular reconstructions of gibbets—such as occur in local ghost walks, computer games and theme parks—often misrepresent the gibbet as a kind of oubliette, where condemned prisoners were left to die of thirst or exposure. There is no evidence that by the eighteenth century this ever happened in Britain. *The Old Englander* reports that in France malefactors might be sentenced to hang in chains for two days *before* execution, being left bareheaded and fed only on bread and water, and then executed on the third day.[72] There are cases of gibbeting alive known from the Caribbean during the plantation period, always in regard to a slave found guilty of a treasonous crime.[73]

Myth 3: There Were Traditional Gibbeting Sites

Many larger towns had a traditional place of execution, especially those in which assizes were held, usually on land close to the county gaol. Larger cities might have a permanent gallows, although several larger towns, including Bath for example, did not have any traditional place of execution. Gibbet locations, as opposed to scaffolds for execution, were generally determined by other factors such as proximity to the scene of the crime, public visibility and the ease of maintaining public order in the large crowds that often attended a gibbeting.

Myth 4: Gibbets Were Occupied by a Series of Bodies

Some of the gibbets used by the Admiralty courts seem to have occupied customary locations and to have hosted a series of bodies. The gibbet cage now in possession of the London Docklands museum, which was

---

[72] *Old Englander*, 25 January 1752.
[73] William Beckford, *Remarks Upon the Situation of the Negroes in Jamaica* (London, 1788), 93; Trevor Burnard *Mastery, Tyranny, and Desire: Thomas Thistlewood and His Slaves in the Anglo-Jamaican World* (Chapel Hill: University of North Carolina Press, 2004), p. 151.

almost certainly an Admiralty one, shows signs of repair which would be redundant on a single-use artefact. However, most cases of hanging in chains as a result of sentences passed by the assize courts involved making a special gibbet-cage fitted to a single individual which then stayed *in situ* with the remains of that particular criminal until the gibbet finally fell or was removed, which was often many decades later (see discussion in Chap. 3). Gibbet irons were not normally reused. This made the costs of gibbeting a single individual very high. The details of exactly how and where a gibbeting took place are considered further in the next chapter.

Open Access This chapter is licensed under the terms of the Creative Commons Attribution 4.0 International License (http://creativecommons.org/licenses/by/4.0/), which permits use, sharing, adaptation, distribution and reproduction in any medium or format, as long as you give appropriate credit to the original author(s) and the source, provide a link to the Creative Commons license and indicate if changes were made.

The images or other third party material in this chapter are included in the chapter's Creative Commons license, unless indicated otherwise in a credit line to the material. If material is not included in the chapter's Creative Commons license and your intended use is not permitted by statutory regulation or exceeds the permitted use, you will need to obtain permission directly from the copyright holder.

CHAPTER 2

How to Hang in Chains: How, Where and When Eighteenth-Century Sheriffs Organised a Gibbeting

Abstract The criminal corpse undertook a journey from the scaffold to the gibbet. The gibbet was commonly located near the scene of crime and in a conspicuous location, usually within sight of a major road. Customary gibbet places existed in London and in some coastal location, but usually the body was transported from the place of execution to the place of hanging in chains. Sometimes, especially earlier in our period, criminals were executed and hung in chains from the same scaffold at the scene of crime. Gibbet cages were made quickly and did not develop local styles. The scene of a gibbeting was often a rowdy and carnivalesque occasion.

Keywords Gibbet · Landscape · Technology · Location · Carnival

THE PROCESS

From the Scaffold to the Gibbet

The progress of the body from the scaffold where execution had occurred to suspension in a gibbet cage typically involved several stages. After execution, the body was left hanging for up to about an hour, both to ensure that there was no sign of life remaining (although recent research by Elizabeth Hurren suggests that in a substantial minority of cases even after such a long period of hanging medical death had not taken place by the time the body was removed from the scaffold) and to allow the crowds who came to view the execution enough time to inspect the body, after which it was taken down and removed to some place where it could be prepared for suspension and enclosed in irons.[1] In some cases, there was a further opportunity to display the new corpse before the gibbeting, with financial benefits for those of an entrepreneurial bent. After the execution of Robert Carleton at Diss, Norfolk, in 1742, for example, his body was carried back to the house where the murder was committed and "hung up upon a balk in the middle of the room, and shewn at two pence a piece. The following day his body was put into its gibbet and displayed at Diss common".[2] Carleton's case was especially salacious, as he was found guilty of the murder of his male lover's wife.

There is little evidence about how the body was dressed for gibbeting. According to newspaper reports, the body of James Cook (d. 1832) was dressed again in the clothes in which he had been executed—probably his best clothes. The sheriffs' cravings for Shropshire 1759 itemise the costs of "plank cords and hair cloth to inclose the bodies" of two men between execution and gibbeting.

Several newspaper accounts of the preparation of the body mention that the corpse was 'tarred' or 'dipped in tar' before being gibbeted. No soft tissue of a gibbeted body survives to allow us to test this, although tarring is frequently mentioned in secondary sources, usually without additional evidence. Neither tar nor anything like it is ever itemised in the sheriffs' cravings relating to gibbetings, despite the separate listing of other apparently trivial expenses such as the cost of ale for guards, rope for a noose, or a stool for a burning. 'Dipping the body in tar' is mentioned in a few of the later newspaper accounts, such as the account of the execution and display of James Cook in 1832. It is possible that tar was used only very occasionally, despite a popular belief that tarring was a normal part of the process. Moreover, it is not clear what this 'tar' might be: if used, it is unlikely to have been a very

---

[1] Elizabeth Hurren (2015) *Dissecting the Criminal Corpse: Post-Execution Punishment from the Murder Act (1752) to the Anatomy Act (1832)* (Basingstoke: Palgrave).

[2] D. Stoker 1990. 'The tailor of Diss: sodomy and murder in a Norfolk town'. Paper published online at http://users.aber.ac.uk/das/texts/tailor_of_diss.htm, accessed 8/7/15.

heavy caulking bitumen which would obscure the individual identity of the body beyond recognition, and one can only imagine that it would make dressing the corpse and enclosing it in its gibbet cage a very difficult and sticky business. James Cook's body was dressed again in his normal clothes following tarring, so the process probably left the body more or less the same size and thus is unlikely to involve a thick, viscous or gluey kind of tar. It was certainly flammable, however, if the tale of the Chevin highwaymen is true. Cox recounts a story of no clear date which is at present unsubstantiated in the historical records: three highwaymen were apprehended and condemned to hanging in chains around the middle of the eighteenth century. Their bodies were gibbeted at the top of the Chevin near Belper in Derbyshire. "After the bodies had been hanging there a few weeks, one of the friends of the criminals set fire, at night-time, to the big gibbet that bore all three. The father of our aged informant, and two or three others of the cottagers nearby, seeing a glare of light, went up the hill, and there they saw the sickening spectacle of the three bodies blazing away in the darkness! So thoroughly did the tar aid this cremation, that the next morning only the links of the iron chains remained on the site of the gibbet". A similar story is told of the 'flaming gibbet of Galley Hill' in Bedfordshire, which may relate to the 1744 gibbeting of John Knott, who was hung in chains on 'Luton Down'.[3] Whatever tarring took place, it at best only delayed the normal process of decomposition. An article in the Buckinghamshire Record Office written by J. Wharton in 1860 records that farmers living up the valley from the place where Corbett was gibbeted on Bierton Common were unable to open their windows for about a year because of the smell from the body.[4]

Despite folkloric—and untrue—accounts of live gibbetings, in the eighteenth and nineteenth centuries criminals were always executed before being hung in chains. In the rare cases that a condemned person managed to evade the gallows by taking their own life, hanging in chains

---

[3] J.C. Cox (1890), untitled note *The Antiquary* (November 1890), p. 214. The Galley Hill story is told at http://myths.e2bn.org/mythsandlegends/playstory39-the-flaming-gibbet-of-galley-hill.html, but there are numerous historical errors and no original sources cited in this retelling.

[4] J. Wharton (1863) 'The last gibbet in Buckinghamshire' *Records of Buckinghamshire* (vol. 2).

might be carried out to confirm the death. Joseph Armstrong, sentenced to hang for the murder of his employer's wife in 1777, managed to break his own neck in prison but nevertheless was gibbeted near the home of his victim in Cheltenham "to obviate every doubt that may be raised of his not being dead".[5] This contrasts with the burial at Tewkesbury crossroads of condemned murderer William Birch, following his suicide in prison in 1791. Staked burial of suicides in the road was common for those who could be considered to have evaded punishment by taking their own lives, but gibbeting would accomplish the same goals of keeping the suicide out of consecrated ground and materially confirming their deviant status through non-normative mortuary treatment.

LOCATING A GIBBET: THE MACRO-GEOGRAPHY OF GIBBETING

In most of England, post-mortem punishment was ordered by the circuit judge presiding at the assize court. Nevertheless, there are clear regional differences in the frequency with which gibbeting was ordered. Table 2.1 shows the frequency of gibbetings by county and decade through England and Wales. The data for the period before the Murder Act is less secure than for the later part of the eighteenth century, and numbers for the decades at the beginning of the century are very incomplete.

London (mostly recorded under "Middlesex" in the county table above) had far and away the most gibbetings. This is due in part to the fact that many of the most serious crimes were tried in London, even if committed elsewhere. London also had a huge population and well-known social and economic problems.[6] However, even in the rural provinces, there were marked differences in the frequency of hanging in chains. In the counties of Sussex, Essex, Gloucestershire and Hampshire, for example, five or six gibbetings sometimes occurred within a decade, whereas in Cornwall there were none at all during the whole period and in County Durham there was only one. These figures take no account of the size or population of a county or of the conviction rates for murder and capital crime.

---

[5] N. Darby (2011). *Olde Cotswold Punishments* (Stroud: The History Press), pp. 24–25.

[6] J. White (2012) *London in the eighteenth century: a great and monstrous thing* (London: Bodley Head).

Table 2.1 The frequency of gibbetings by county and decade through England and Wales

	1700*	1710*	1720*	1730*	1740	1750	1760	1770	1780	1790	1800	1810	1820	1830	County total
Beds					1	1									2
Berks				1			1		2		1				5
Bucks				2	1	1		1							5
Caernarfon					2			1							3
Camarths							1		1						2
Cambs					1	1				2					5
Cheshire				2		2		1		4			1		9
Cornwall															0
Cumbria					1		2								3
Denbeighs			1					1				1			3
Derbs								1							2
Devon				2	1	1			4	2					10
Dorset						1	1								2
Durham								1						1	2
Essex				1		6				1					8
Flintshire					1			1							2
Glamorgs			1			2	2	1	3						6
Gloucs			1	2	5	1	2	2	2	5	3				16
Hamps			2	1	1	1	2	2	2	1					19
Herefs															4
Herts					3	2	3		2		1				10
Hunts									1						1
Kent				1	2	3	3		2	1	1				13
Lancs						1	2			2	1				6
Leics											1			1	2
Lincs				2	1	1	2			1	1				8

*Figures for gibbetings before the 1730s are mostly unavailable

(continued)

Table 2.1 (continued)

	1700*	1710*	1720*	1730	1740	1750	1760	1770	1780	1790	1800	1810	1820	1830	County total
Middx	5	5	12	16	13	4	6	3	2	2					68
Mons					1										1
Montgoms				1											1
North'nd				1											2
Norfolk					3	2			2	1					12
Notts							1	1		5					2
Northants				1	1	1	1								4
Oxon					1				1						2
Pembs											1				1
Radnor						1		1							2
Rutland							1		2						3
Salop			2	1	1	2	1			1					6
Somerset			2	2	5			1	3	3		1			12
Staffs				1	2	2	1		2	2					7
Suffolk				3	2	3	1		3	1					10
Surrey			2	2	10										15
Sussex				1			3	1	3	4	1				17
Warks					2		2	2	1						8
Wilts							1	1	1						7
Worcs							2	2	1						3
Yorks						1				1		2	1	2	7
Decade total	5	5	23	44	65	40	41	26	38	39	10	2	1	2	338

The Micro-Geography of Gibbeting

Location of gibbets was specified neither in law nor, usually, in the judge's sentence. Sometimes, the sentence of hanging in chains specified only that the gibbet should be "at a convenient location" and close to the scene of crime. My attempts to locate the original positions of gibbets have mostly been unable to pinpoint exact locations, but I have often been able to identify the place within 50 metres or so because a particular road junction or landmark is mentioned (see map). Newspaper reports are sometimes very specific: William Whittle (executed 1766 in Lancaster) was gibbeted, according to the newspaper, "at the Four Lane Ends, within forty yards of his father-in-law's house, and a hundred yards of his late dwelling house, about three miles from Preston, on the Liverpool road by way of Croston".[7]

There is some geographical variability in the kind of locations that were selected as suitable sites for the erection of a gibbet. London gibbets were erected in customary places on areas of open land; those convicted of maritime crimes might be hung in chains around the coast where they would be visible to shipping. Both of these types of gibbetings are considered below. However, for the majority of those sentenced to gibbeting at provincial assizes, the three most prevalent concerns when selecting a gibbet location are proximity to scene of crime, visibility from the public road, and the capacity of the immediate locale to cope with a large crowd. Although the second of these is rarely specified in either the sentence or the cravings, a study of the locations of gibbets, where known, demonstrates a close correlation between gibbet sites and proximity to what are now A-roads in Britain. A 1755 travellers' guide to London explains to foreign visitors that gibbets were situated on the highway "near the Place where the Fact was done, to perpetuate the Villainy of the Crime, and to serve as an Example".[8] To be effective as a warning or a deterrent, the gibbet had to be highly visible. But this requirement was sometimes in conflict with the needs of travellers to pass freely without being either inconvenienced by crowds of vulgar spectators or brought into such close contact with the decomposing bodies that delicate sensibilities would be offended. The location of John

---

[7] *General Evening Post*, 12–15 April 1766 issue 5069.
[8] Anon (1755) *London in Miniature* (London: C. Corbett), p. 217.

Haines's gibbet on Hounslow Heath, erected in 1799, was criticised in one London newspaper for being "shamefully placed close to the high road", a criticism rejected by another paper, which claimed that in fact "so far from being offensively situated, [it] is placed at the distance of at least five hundred yards from the high road".[9]

When the gibbet was sited, use was sometimes made of natural features in the landscape that enhanced its visibility. When John Naden was hung in chains in Staffordshire in 1731, his gibbet was erected "on the highest hill on Gun Heath" within a quarter mile of the house where he murdered his master.[10] Lingard's gibbet near St Peter's rock in the Derbyshire peaks was sited near the main road and close to the tollbooth whose keeper he had murdered. Its visibility from both the scene of crime and the highway more generally was accentuated by taking advantage of a natural local landmark (Fig. 2.1). The roads along which gibbets were located were also, as far as possible, major routes, and the gibbets were close to junctions where they would be noticeable from at least two roads. Ogilby's seventeenth-century linear maps of British journeys show a number of gibbets, marked as waymarks in the same way that prominent windmills or stands of trees are included. Interestingly, when the 1628 gibbet of John Felton, murderer of the Duke of Buckingham, fell down, it was replaced by an obelisk in 1782—not as a memorial to Felton or the Duke of Buckingham but to serve as a boundary marker as the old Southsea gibbet had come to mark the boundary of the borough of Portsmouth (Fig. 2.2). The obelisk, photographed in the 1930s, has since disappeared.[11]

In this period, however, there is comparatively little evidence of the re-use of archaeological sites for gibbets, in contrast to widespread medieval practice and practice on the continent in early modernity.[12]

---

[9] *Whitehall Evening Post*, 12–14 March 1799, issue 8056; *Morning Herald*, 15 March 1799, issue 5769.

[10] *Daily Courant*, 13 September 1731.

[11] www.memorials.inportsmouth.co.uk/southsea/obelisk.htm.

[12] See, for example, H. Williams (2006) *Death and Memory in Early Medieval Britain* (Cambridge: Cambridge University Press); Andrew Reynolds (2009) *Anglo-Saxon deviant burial customs* (Oxford: Oxford University Press); J. Coolen (2014) 'Places of justice and awe: the topography of gibbets and gallows in medieval and early modern north-western and Central Europe' *World Archaeology* 45(5), pp. 762–79.

Fig. 2.1 St Peter's rock, Derbyshire, where Anthony Lingard was hung in chains in 1815 (photo: Sarah Tarlow)

John Nichols, whose gibbet cage was excavated in the twentieth century and now is on display in Moyses Hall, Bury St Edmunds, was originally hung in chains on a prehistoric burial mound called Troston Mount at Honington in Suffolk, and Michael Morey's Hump on the Isle of Wight is now named for the murderer gibbeted there in 1737 but is in fact a Bronze Age mound. Combe gibbet, erected in 1676, makes ostentatious use of an earlier monument—the Inkpen neolithic long barrow on Gallows Down—and according to most stories of that double gibbet, it was placed on a parish boundary in order that the costs might be shared

Fig. 2.2 Felton's obelisk in Portsmouth. (Portsmouth: Charpentier)

The Obelisk which formerly enclosed the gibbet upon which Felton's body was hung. It also marked the east point of the Portsmouth Boundary between the two adjacent parishes to which the two convicted murderers belonged, so the mound's function as a parish boundary marker might be more significant than its archaeological heritage. Whyte suggests in her Norfolk-based study that gibbets were usually sited on or close to parish boundaries, where they crossed open or common land, a location which had a symbolic importance of considerable time depth. She traces the use of boundary points for criminal execution and burial to early medieval beliefs about the liminality of the dangerous and deviant dead, through the traditional use of boundaries for siting gallows by the manorial courts of the later middle ages. Whyte argues that, by the later medieval period, the placement of gallows at parish boundaries "was intended not so much to convey any idea of territorial 'marginality', but rather to denote the eviction of the condemned from the spiritual core of the community—represented in the landscape by the church and the graveyard".[13] My research has not found the close relationship between gibbet locations and parish boundaries to be so evident elsewhere; the twin principles of siting the gibbet close to a public road and near the scene of crime, however, do produce a gibbeted landscape of marginal land—commons, verges, heaths and forests, which might increase the chances that a gibbet would be sited on or close to a parish boundary. The interpretation that gibbet locations are geographical representations of the criminal's exclusion from normal society is strong and, though probably not the primary determinant of siting the gibbet, undoubtedly added to the force of witnessing it.

The likelihood that large crowds of people would attend the gibbeting militated towards the selection of open or common land as gibbet sites, but on other occasions private land was used. The sheriffs' cravings for Devon in 1752 note a cost incurred "for an express to go to Mr. Wiscotts about 40 miles to ask leave to erect a gibbet on his manor in order to hang up the body of John Young (not receiving the order for so doing til he was executed)".

As conspicuous locations in public places, gibbets acted as meeting points and landmarks. Jeremiah Abershaw's gibbet on Putney Heath is mentioned in the press as the location of a boxing match in 1796 and 1800, a duel in 1798, and a military tattoo in 1803. In 1773, at Kennington Common

---

[13] N. Whyte (2003) 'The deviant dead in the Norfolk landscape' *Landscapes* 1, p. 35.

Several gentlemen, frequenters of a very genteel public house not far from Duke's Court, Bow Street, Covent Garden, intend in a few days to decide a wager depending on a game at Trap Ball on Kennington Common in a very whimsical and humorous manner; three fourths of the party being either old, lame with gout corns or other inflammations affecting feet and legs, while the rest are young, nimble and alert, have unanimously agreed to play the game each person in a wheel barrow, which is managed by the strongest and most fore footed Irish chairman anywhere to be found. As this expedient puts all parties on equality, it is expected there will be much sport and fair play unless charioteers should be bribed to jockey one another. The parties to rendezvous as near the gibbets as possible.[14]

One of the remarkable features of gibbet locations is the apparent irrelevance of buildings or monuments that signify State power. Laqueur has noted that many eighteenth- and nineteenth-century executions did not take place close to the civic or judicial apparatus of State or local authority but rather customarily used a suburban location of little distinction.[15] Similarly, the places where the body was fixed for display were very rarely in urban locations and were usually far from any building or monument meaningful to the working of the institutions of State, although they were rarely far from roads.

Proximity to the scene of crime often also resulted in proximity to the home of the criminal, although there is no evidence that gibbets were deliberately sited to be close to the home of either the criminal or the victim. Such siting, however, could have harsh consequences for the family of the gibbeted man. Not only would they be confronted regularly with the horrible spectacle of the body of their relative in the process of decay which would be emotionally upsetting, their friends and neighbours would also be constantly reminded of the evil done by their kin. This was the reason that Thomas Willdey's family petitioned the sheriff in 1734 asking for his body to be removed from its site on Witley Common near Coventry: the innocent and respectable members of the family were subject to comment, abuse and loss of trade in their local

---

[14] *Middlesex Journal or Universal Evening Post*, 22–24 July 1773, issue 674.

[15] T. Laqueur (1989) 'Crowds, carnivals and the State in English executions, 1604–1868', in A.L. Beier, D. Cannadine, J.M. Rosenheim (eds.) *The First Modern Society* (Cambridge: Cambridge University Press), pp. 305–55, p. 312.

area by reason of the continuing presence of Willdey's "offensive" body.[16] Thomas Jackson, convicted of robbing the mail, was hung in chains on Methwold common "near the place where he committed the robbery", which was "directly opposite to the dwelling of his unfortunate family".[17] In 1797, an estate map of the village of Nether (or Lower) Hambleton in Rutland, now drowned beneath Rutland Water, annotated one cottage with a cross and the note "where you never go—mother to the young men that were hanged".[18] The young men in question were the Weldon brothers, gibbeted for murder within sight of their parents' cottage. One can only imagine the difficulties of living on in a small community and enduring every day the sight of your sons' dead bodies, as well as what sounds like ostracism by your neighbours.

Hanging at the Scene of Crime

Gibbets were usually erected for the display of the dead body only; the criminal had actually died on a scaffold constructed at the customary place of execution for that town.

Steve Poole has studied the incidence of scene-of-crime executions in Britain.[19] Between 1720 and 1830, at least 211 people were hanged on specially erected scaffolds at the scene of their crime. More than half of these crime-scene hangings took place in the southeast (London and Surrey) and in Gloucestershire and Somerset. Most of those executed at the scene of crime were taken down and disposed of elsewhere, but a minority were subsequently enclosed in gibbet cages and then hung up again on the same framework. Poole suggests that although the sheriff

---

[16] TNA SP 36/32/115.

[17] *London Chronicle*, 1–3 April 1790, issue 5244; *Public Advertiser*, 17 April 1790, issue 17403.

[18] S. Sleath and R. Ovens (2007) 'Lower Hambleton in 1797', in R. Ovens and S. Sleath (eds.) *The Heritage of Rutland Water* (Rutland Record Series number 5) Oakham: Rutland Local History and Record Society, pp. 193–209.

[19] Steve Poole (2015) 'For the Benefit of Example': Crime-Scene Executions in England, 1720–1830', in R. Ward (ed.) *A Global History of Execution and the Criminal Corpse*. Basingstoke: Palgrave: 71–101; Steve Poole (2008) 'A lasting and salutary warning': incendiarism, rural order and England's last scene of crime execution', *Rural History* 19, pp. 163–77.

rarely made explicit the reason for holding an execution at the scene of the crime instead of in the customary location, crimes of brutal murder and crimes involving foreigners or where there was a high risk of crowd disorder were most likely to be singled out in this way. Scene-of-crime executions were highly personalised and related with a pleasing symmetry to the crime, which might have helped to satisfy the popular appetite for balanced revenge.[20] Demonstrations of State power in out-of-the-way spots also helped, argues Poole, to re-establish local authority in isolated rural areas. Contemporary commentators were also impressed by the sentimental potential of the malefactor's last moments incorporating a view of his childhood haunts and the place of his undoing. A powerful dramatic experience such as a scaffold confession or visible moment of contrition would be more readily provoked and make a more emotionally powerful impression in this highly theatrical setting.

Gibbets in the Landscape

The presence of a gibbet could change the experience of a local landscape for a long time after its erection, even to the present day. The large crowds and carnival atmosphere of the newly erected gibbet would continue for only a few weeks, but the memory of the unusual event would last a lifetime for those who had been present. Moreover, the gibbet itself often remained standing for many decades and would affect both the experience of travelling through the landscape and the way in which the landscape was known. Gibbet locations and former gibbet locations acted as landmarks. The inclusion of eight gibbets on Faden's 1797 map of Norfolk demonstrates that the gibbets were important landmarks. Earlier national maps, such as Ogilby's *Britannia* of 1675, show several gibbets along with windmills, bridges and other fixed points by which the progress of a road journey would be marked.

A gibbet might remain in place for many decades. Since they were often ten metres or more in height, they were conspicuous in the landscape and affected the way that local people knew and experienced the area, through giving new names to the roads and fields where they were sited, and by giving emotional impact to local journeys, or even motivating the creation of new routes. Ralph's Lane and Tom Otter's Lane in Lincolnshire, Old

---

[20] Public discussion at the time of the Murder Act included a number of voices in favour of some sort of *lex talionis*—a punishment regime which mirrors the nature of the crime.

2 HOW TO HANG IN CHAINS: HOW, WHERE AND WHEN ... 47

Fig. 2.3 Road sign, Gibbet Hill Lane, Scrooby (photo: Sarah Tarlow)

Parr Road in Banbury and Curry's Point near Whitley Bay are among the places named after the criminals who were gibbeted there. The numerous instances of 'Gibbet Woods', 'Gibbet Hill' and 'Gibbet Lane' in England are hard to date and in most cases are probably medieval in origin, but a number of those are associated with known eighteenth-century gibbets, such as the Gibbet Hill Lane at Scrooby, close to the 1779 gibbet of John Spencer (Fig. 2.3). What was it like to travel through a landscape populated with the remains of the dead? Sources suggest that, for most people who lived close to these structures or encountered them regularly, it was at best distasteful and often quite horrifying and that people would take measures to avoid passing a gibbet when possible, especially at night. A report in the Buckinghamshire archives notes that

"the footpath running from 'Chalkhouse Arms' and continuing back along the back of the hovels in Bierton, as far as 'the milestone' dates from this execution, and was made in order to avoid passing the gibbet [of Corbett, d. 1773]".[21] W.H.B. Sanders[22] says that the former ostler at a nearby inn recalled seeing Gervase Matcham's gibbet on rough ground adjacent to the Great North Road:

> It often used to frit me as a lad. I have seen horses frit with it. The coach and carriage people were always on the look out for it. Oh yes! I can remember it rotting away, bit by bit, and the red rags flapping from it. After a while they took it down and very pleased I were to see the last of it.

The ostler's revulsion was shared by the young Charlotte Latham, who remembered the full sensory assault of having to pass a gibbet on the Brighton Road in her childhood: "Standing on the wide desolate down, with all its fearful associations, it was an object of great terror to me in my youthful days; and the dread of seeing it and hearing my nurse repeat her oft-told tale of the murderer who had been hung on it in chains, and how he had been swinging on a windy night and heard rattling his irons, made the prospect of a visit to the sea-side, which involved the sight of the gallows anything but pleasurable".[23]

Wordsworth famously remembered his encounter with the site of Thomas Nicholson's gibbet at Penrith. Nicholson had been gibbeted in 1767, and by the time Wordsworth came to the spot
"The gibbet-mast had mouldered down, the bones
And iron case were gone".[24]

However, as Duncan Wu notes, if Wordsworth is reminiscing about the year 1775, the gibbet mast would not have mouldered down and indeed a five-year-old child would probably not have ridden so far unaccompanied. Wu suggests that another Cumbrian gibbet may have been intended.[25] Whatever the case, *The Prelude* is not an accurate historical record but, for our present purposes, a good indicator of the response of a sensitive Romantic spirit to the presence of gibbets in the landscape: Wordsworth "fled/Faltering and fain, and Ignorant of the road".

---

[21] From a letter dated April 18 1860, signed by J. Wharton to Rev. C. Lowndes. At http://myweb.tiscali.co.uk/corbettonenamestudy/First/Books/Extract2.htm.
[22] W.H.B. Sanders (1887) *Legends and traditions of Huntingdon* (London: Simpkin, Marshall and Co.), pp. 103–04.
[23] Charlotte Latham (1868) *Some West Sussex superstitions lingering in 1868, collected by Charlotte Latham at Fittleworth* (London: The Folk-Lore Society).
[24] W. Wordsworth *The Prelude* 1805 version 11: lines 290–01; lines 299–300.
[25] D. Wu (2002) *Wordsworth: An Inner Life* (Oxford: Blackwell), p. 465.

When Mary Hardy, a Norfolk farmer's wife, visited North Yorkshire, she made a special trip to see the gibbet of Eugene Aram in Knaresborough 16 years after it had been erected there.[26]

Out of the Ordinary

Most gibbets were erected in rural areas by county sheriffs in direct response to an assize court judgement of a single person. There were, however, two variants of hanging in chains which followed different customs—those in London and those carried out by the Admiralty courts. Let us look briefly at both.

Exception 1: London
London was an exceptional city throughout the period of study, as it has been throughout post-Roman British history and as it remains today. Those convicted of a capital offence in London were more likely to die than those convicted elsewhere in Britain (where pardons were common or a non-capital sentence was substituted).[27] Although London had only 10% of the population, it produced 30% of the executed bodies. This exceptionalism also affected the location of London gibbets. Whereas in most of Britain gibbets were erected for their proximity to the scene of crime and for visibility from the main road, those sentenced to hanging in chains in the metropolis were not put close to the place of their crimes but in one of a small number of traditional gibbet locations. These were usually pieces of open land just outside the city but adjacent to one of the main roads in and out of London. Finchley Common, Hounslow Heath, Bow Common and Shepherd's Bush were all areas used for several gibbets, and a number of others were erected along the Edgeware Road. Although these locations did not relate to the details of the crime or indeed to the criminal's biography in any way, they were well chosen for public visibility and, for the most part, permitted the formation of large gibbet crowds without threatening public order. The reason for siting London gibbets on one of a few regularly used open locations rather

---

[26] *Mary Hardy's Diary* (ed. B. Cozens-Hardy). 1938. Norfolk Record Society Vol. 37.
[27] P. King and R. Ward (2016) 'Rethinking the Bloody Code in Eighteenth-Centre Britain: Capital Punishment at the Centre and on the Periphery.' *Past and Present* (2016).

than at the scene of crime is not discursively addressed in contemporary literature, but it is likely that the dense urban landscape in which most of the crimes took place was impractical for erecting gibbets. Although the streets and squares of London hosted various carnivals, markets and events, these were ephemeral events. Gibbets normally remained standing for decades. The large crowds drawn to a gibbeting could not be accommodated in an orderly way in the narrow roads of the capital, where traffic would be stopped and the risk of public disorder was always high. Moreover, the continuing presence of a rotting corpse among the dense habitations of the living would surely have been considered unpleasant even before the hygienic reforms of the mid-nineteenth century would have condemned it as unsanitary. Early nineteenth-century reformers campaigning for the closure of overfull urban graveyards emphatically condemned the proximity of dead bodies to the places of the living. The 'miasmas' of infection produced by the decaying body had injurious or fatal consequences for the health of the living.[28]

Exception 2: The Admiralty Courts and Maritime Crimes
In addition to those criminals who passed through the normal assize or London courts, at least 87 men were sentenced to death by the Admiralty courts between 1726 and 1830, and the records of the Admiralty court enable some analysis of this group (Table 2.2). The Admiralty courts dealt with crimes committed at sea and were mostly for murder at sea or piracy. Mutiny, theft and sinking or destroying a ship also resulted in a small number of capital convictions. Of these 87 men, at least 38 were gibbeted, at least 3 were dissected, 6 were interred without further punishment, 3 had their sentences commuted to transportation, one was rescued from the scaffold, and the fate of the others is unknown, although most of them were sentenced to dissection. We have not been able to trace the ultimate fate of some of those sentenced. Even if we assume that none of those whose ultimate fate is unknown was gibbeted, 38 out of 87 is a very high proportion—much higher than the proportion of murderers sentenced to be hung in chains by the terrestrial courts. Hanging in chains for those convicted by the Admiralty Court was distinctive in a number of ways. First, the Admiralty courts made repeated use of customary locations for both execution and gibbeting. Most executions were

---

[28] See, for example, G. Walker (1839) *Gatherings from Graveyards* (London: Longman).

2 HOW TO HANG IN CHAINS: HOW, WHERE AND WHEN … 51

Table 2.2 Admiralty Court convictions resulting in hanging in chains

Year of execution	Name	Crime	Date of sentence	Execution date	Sentence	Carried out?	Where?	Notes
1726	John Jean (Captain Jane)	Murder		Friday, 13 May	Unknown	Gibbeted	Greenwich	Gale/ECCO
1726	John Gow	Piracy and Murder		11 June 1726	Unknown	Gibbeted	Greenwich	Gale/ECCO
1726	James William	Piracy and Murder		11 June 1726	Unknown	Gibbeted	Blackwall	Gale/ECCO
1735	Thomas Williams	Murder of James Beard		Friday, 14 March	Unknown	Unknown		Gale
1737	Richard Coyle	Murder of Benjamin Hartley		Monday, 14 March	Unknown	Gibbeted	Gallions beyon Erith	OBO, Gale
1737	Edward Johnson	Murder and Piracy		Monday, 14 March	Unknown	Gibbeted	Half Way Tree	OBO, Gale
1737	Nicholas Williams	Murder and Piracy		Monday, 14 March	Unknown	Gibbeted	Half Way Tree	OBO, Gale
1737	Lawrence Sennett	Murder and Piracy		Monday, 14 March	Unknown	Gibbeted	Gallions beyon Erith	OBO, Gale
1738	John Richardson	Murder of Benjamin Hartley		Saturday, 28 January	Unknown	Unknown		OBO, Tried with Richard Coyle
1738	James Buchanan	Murder of Michael Smith		Wednesday, 20 December	Unknown	Carried away	Possibly still alive, escaped abroad	OBO, Gale
1743	Thomas Rounce	High Treason (fighting on a Spanish privateer)		Wednesday, 19 January	Unknown	Interred	By friends	Gale
1744	Andrew Miller	Murder of James Nelson	December 1743	Monday, 21 February	Unknown	Unknown		OBO, Gale

(continued)

Table 2.2 (continued)

Year of execution	Name	Crime	Date of sentence	Execution date	Sentence	Carried out?	Where?	Notes
1752	Captain James Lowry	Murder of Kenneth Hossack		Wednesday, 25 March	Unknown	Gibbeted	Galleons below Woolwich	ECCO, Gale
1754	John Lancey	Destroyed the Nightingale		Friday, 7 June	Unknown	Interred		Gale
1759	Joseph Halsey	Murder of John Edwards		Wednesday, 14 March	Dissection	Unknown		Gale
1759	Nicholas Wingfield	Piracy		Wednesday, 28 March	Unknown	Gibbeted	Captain Lowry's Gibbet	Gale
1759	Thomas Hide	Piracy		Wednesday, 28 March	Unknown	Gibbeted	Blackwall Bridge	Gale
1759	William Lawrence	Piracy	Monday, 29 October	Wednesday, 19 December	Death	Unknown	Unknown	HCA 1/61
1760	John Tune	Piracy	28 March?	Monday, 8 December	Death	Gibbeted	Unknown	HCA 1/61 (OBO)
1762	Thomas Smith	Piracy	Tuesday, 30 March	Monday, 10 May	Death	Gibbeted	Blackwall	HCA 1/61, Gale
1762	Robert Main	Piracy	Tuesday, 30 March	Monday, 10 May	Death	Gibbeted	Blackwall	HCA 1/61, Gale
1767	John Winne	Murder of a negro sailor	Friday, 7 February	Tuesday, 10 March	Dissection	Unknown	Surgeons Hall	HCA 1/61, HCA 1/87
1769	Thomas Ailesbury	Piracy and robbery	Tuesday, 31 October	Wednesday, 29 November	Death	Gibbeted	Unknown	HCA 1/61, HCA 1/87, GALE

(continued)

Table 2.2 (continued)

Year of execution	Name	Crime	Date of sentence	Execution date	Sentence	Carried out?	Where?	Notes
1769	Samuel Ailsbury	Piracy and robbery	Tuesday, 31 October	Wednesday, 29 November	Death	Gibbeted	Unknown	HCA 1/61, HCA 1/87, GALE
1769	James Hide	Piracy and robbery	Tuesday, 31 October	Wednesday, 29 November	Death	Unknown		HCA 1/61, HCA 1/87, GALE
1769	William Geary	Piracy and robbery	Tuesday, 31 October	Wednesday, 29 November	Death	Unknown		HCA 1/61, HCA 1/87, GALE
1769	William Wenham	Piracy and robbery	Tuesday, 31 October	Wednesday, 29 November	Death	Unknown		HCA 1/61, HCA 1/87, GALE
1769	Edward Pinnell	Sinking a Ship	Tuesday, 31 October	Wednesday, 29 November	Death	Unknown		HCA 1/61, HCA 1/87, GALE
1772	David Ferguson	Murder of Peter Thomas	Tuesday, 18 December 1771	Thursday, 3 January	Gibbeting/ dissection	Gibbeted	Marshes by the river	HCA 1/61, HCA 1/87
1772	John Shoales	Piracy	7 November 1771	Wednesday, 11 December	Death	Unknown		HCA 1/61, HCA 1/87
1775	Thomas Sawyer	Murder of William Barbet	Wednesday, 8 November	Saturday, 14 November 1775	Gibbeting	Unknown		HCA 1/61
1781	William Townsend	Murder and piracy	Wednesday, 31 October	Saturday, 17 November	Death	Dissection		HCA 1/61, HCA 1/87
1781	James Sweetman	Felony and Piracy	Wednesday, 31 October	Tu. 4 December	Gibbeting	Gibbeted	Kent Coast	HCA 1/85, HCA 1/87, Gale

(continued)

54 S. TARLOW

Table 2.2 (continued)

Year of execution	Name	Crime	Date of sentence	Execution date	Sentence	Carried out?	Where?	Notes
1781	Matt Knight	Felony and Piracy	Wednesday, 31 October	Tu. 4 December	Gibbeting	Gibbeted	Kent Coast	HCA 1/85, HCA 1/87 Gale
1781	William Paine	Felony and Piracy	Wednesday, 31 October	Tu. 4 December	Gibbeting	Gibbeted	Norfolk Coast	HCA 1/85, HCA 1/87 Gale
1784	Samuel Harris	Murder of John M'Neir	Thursday, 11 November	Saturday, 13 November	Death	Gibbeted	Blackwall	HCA 1/61, HCA 1/87, GALE
1784	John North	Murder of John M'Neir	Thursday, 11 November	Saturday, 13 November	Death	Gibbeted	Blackwall	HCA 1/61, HCA 1/87, GALE
1786	George Combes	Murder of William Allen	Saturday, 21 January	Monday, 23 January	Death	Gibbeted	Broad Ness Point	HCA 1/61 Gale
1786	William Hines	Piracy	Saturday, 21 January	W/Thursday, 15/16 Feb.	Death	Gibbeted	Broad Ness Point	HCA 1/61 Gale
1788	Henry Parsons	Mutiny on the "Ranger"	Monday, 12 November 1797	Monday, 14 January	Gibbeting	Interred		HCA 1/85 Gale
1788	George Steward	Mutiny on the "Ranger"	Monday, 12 November 1797	Monday, 14 January	Gibbeting	Interred		HCA 1/85 Gale
1788	Thomas Johnson	Piracy against "La Pourvoyeuse"	Monday, 12 November 1797	Monday, 14 January	Gibbeting	Gibbeted	Unknown	HCA 1/85 Gale
1788	John Ross	Piracy against "La Pourvoyeuse"	Monday, 12 November 1797	Monday, 14 January	Gibbeting	Interred		HCA 1/85 Gale

(continued)

Table 2.2 (continued)

Year of execution	Name	Crime	Date of sentence	Execution date	Sentence	Carried out?	Where?	Notes
1788	John Thompson	Piracy against "La Pourvoyeuse"	Monday, 12 November 1797	Monday, 14 January	Gibbeting	Gibbeted	Unknown	HCA 1/85 Gale
1790	Thomas Brett	Stole off the "Dutch Hoy"	Monday, 30 November 1789	Monday, 4 January	Death	Gibbeted	Between Greenwich and Erith	HCA 1/61, HCA 1/85, Gale
	John Williams	Mutiny on the "Gregson"	Monday, 30 November 1789	Monday, 4 January	Death	Gibbeted	Between Greenwich and Erith	HCA 1/61, HCA 1/85, Gale
	Hugh Wilson	Mutiny on the "Gregson"	Monday, 30 November 1789	Monday, 4 January	Death	Gibbeted	Between Greenwich and Erith	HCA 1/61, HCA 1/85, Gale
	John Clark	Stole off the "Lands End"	Monday, 30 November 1789	Monday, 4 January	Death	Gibbeted	Between Greenwich and Erith	HCA 1/61, HCA 1/85, Gale
	Edward Hobbins	Stole off the "Lands End"	Monday, 30 November 1789	Monday, 4 January	Death	Gibbeted	Between Greenwich and Erith	HCA 1/61, HCA 1/85, Gale
1792	George Hindmarsh	Murder of Samuel Cowie	Friday, 8 June 1792	Friday, 6 July	Dissection	Dissection	Surgeons Hall	HCA 1/61, HCA 1/85
	Charles Berry	Piracy and Felony on the Fawy	Friday, 8 June 1792	Friday, 6 July	Death	Gibbeted	Blackwall Bridge	HCA 1/61, HCA 1/85
	John Slacke	Piracy and Felony on the Fawy	Friday, 8 June 1792	Friday, 6 July	Death	Gibbeted	Blackwall Bridge	HCA 1/61, HCA 1/85
1796	Francis Cole	Murder of William Little	Friday, 22 January	Thursday, 28 January	Dissection	Unknown	Surgeons Hall	HCA 1/61, HCA 1/85
	Michael Blanche	Murder of William Little	Friday, 22 January	Thursday, 28 January	Dissection	Unknown	Surgeons Hall	HCA 1/61, HCA 1/85

(continued)

Table 2.2 (continued)

Year of execution	Name	Crime	Date of sentence	Execution date	Sentence	Carried out?	Where?	Notes
	George Colley	Murder of William Little	Friday, 22 January	Thursday, 28 January	Dissection	Unknown	Surgeons Hall	HCA 1/61, HCA 1/85
1798	George Jay	Felony and Piracy	Monday, 1 December 1797	Monday 5 March 1798	Death	Unknown		HCA 1/61, HCA 1/85
1799	Jean Prevost	Murder of James Wilcox	Friday, 20 December	Monday, 23 December	Dissection	Unknown	Surgeons Hall	HCA 1/61, HCA 1/85
1800	James Wilson	Felony and Piracy	Wednesday, 11 June	9th July	Death	Unknown		HCA 1/61, HCA 1/85
1800	Thomas Potter (25)	Murder	Wednesday, 10 December	Thursday, 18 December	Dissection	Unknown		HCA 1/61, HCA 1/85
1802	William Codling (45)	Destroying the "Adventure"	Tuesday, 26 October	Saturday, 27 November	Death	Interred	Delivered to friends	HCA 1/61, HCA 1/85, HCA 1/112
1806	Andrew Akow (36)	Murder on the high seas	Friday, 11 July	Friday, 18 July	Dissection	Dissected	College, 32 Dukes Street	HCA 1/61, HCA 1/85, HCA 1/112
1809	Capt. John Sutherland (46)	Murder at sea of Richard Wilson	Friday, 23 June	Thursday, 29 June	Dissection	Unknown		HCA 1/61, HCA 1/85
1812	Charles Palm	Murder on the high seas	Friday, 18 December 1812	Monday, 21 December	Dissection	Unknown		HCA 1/61, HCA 1/85
1812	Samuel Telling	Murder on the high seas	Friday, 18 December 1812	Monday, 21 December	Dissection	Unknown		HCA 1/61, HCA 1/85
1812	Thomas Young Husband	Piracy	Friday, 18 December 1812	Monday, 21 December	Death	Unknown		HCA 1/87

(continued)

Table 2.2 (continued)

Year of execution	Name	Crime	Date of sentence	Execution date	Sentence	Carried out?	Where?	Notes
1812	William Jennott	Felony and Piracy	Friday, 28 February	N/A	Death	Transportation		HCA 1/87
1813	John Bruce	Murder on the high seas	Wednesday, 16 December 1812	Monday, 4 January	Dissection	Unknown		HCA 1/61, HCA 1/85
1813	John Wiltshire	Felony and Piracy	Monday, 5 July	Friday, 30 July	Death	Unknown		HCA 1/61, HCA 1/85
1814	Martin Hogan	Murder	Friday, 21 January	Monday, 24 January	Dissection	Unknown		HCA 1/61, HCA 1/85
1814	Manuel Amarro	Stabbing Mark Kirby	Friday, 1 July	?	Guilty	Unknown		HCA 1/87
1814	Pootoo	Murder on the high seas	Tuesday, 13 December	Thursday, 15 December	Dissection then Gibbeting	Gibbeted	Essex/Kent Coast	HCA 1/61, HCA 1/85, HCA 1/112
1814	Moodie (14)	Murder on the high seas	Tuesday, 13 December	Thursday, 15 December	Dissection then Gibbeting	Gibbeted	Essex/Kent Coast	HCA 1/61, HCA 1/85, HCA 1/112
1814	Sootoe	Murder on the high seas	Tuesday, 13 December	Thursday, 15 December	Dissection then Gibbeting	Gibbeted	Essex/Kent Coast	HCA 1/61, HCA 1/85, HCA 1/112
1814	Cadern	Murder on the high seas	Tuesday, 13 December	Thursday, 15 December	Dissection then Gibbeting	Gibbeted	Essex/Kent Coast	HCA 1/61, HCA 1/85, HCA 1/112
1816	John Gillam	Murder	Monday, 22 January	Tuesday, 30 January	Dissection then Gibbeting	Both?	Kent Coast	HCA 1/61, HCA 1/85
1816	William Brockman (40)	Murder	Monday, 22 January	Tuesday, 30 January	Dissection then Gibbeting	Both?	Kent Coast	HCA 1/61, HCA 1/85

(continued)

58 S. TARLOW

Table 2.2 (continued)

Year of execution	Name	Crime	Date of sentence	Execution date	Sentence	Carried out?	Where?	Notes
1816	Robert Smith	Murder at sea	Monday, 18 November	Thursday, 21 November	Dissection/Gibbeting	Unknown	Unknown	HCA 1/61, HCA 1/85
1816	Charles Furney	Murder at sea	Monday, 18 November	Thursday, 21 November	Dissection/Gibbeting	Unknown	Unknown	HCA 1/61, HCA 1/85
1817	John Pierie	Piracy	Monday, 18 November	Tuesday, 7 January	Death	Unknown		HCA 1/61, HCA 1/85
1817	Jonas Norburgh	Piracy	Monday, 18 November	Tuesday, 7 January	Death	Unknown		HCA 1/61, HCA 1/85
1817	Daniel Brace	Piracy	Monday, 18 November	Tuesday, 7 January	Death	Unknown		HCA 1/61, HCA 1/85
1817	William Hastings	Piracy	Monday, 18 November	Tuesday, 7 January	Death	Unknown		HCA 1/61, HCA 1/85
1820	James Pater	Murder at sea	Friday, 28 January	Tuesday, 1 February	Death	Unknown		HCA 1/61
1830	William Watts	Piracy	Thursday, 4 November	Thursday, 9 December	Death	Unknown		HCA 1/111, HCA 1/88
1830	George Davis	Piracy	Thursday, 4 November	Thursday, 9 December	Death	Unknown		HCA 1/111, HCA 1/88
1830	William Stevenson	Piracy	Thursday, 4 November	N/A	Death (commuted)	Transportation		HCA 1/111, HCA 1/88
1830	Beveridge	Piracy	Thursday, 4 November	N/A	Death (commuted)	Transportation		HCA 1/111, HCA 1/88

ECCO, Eighteenth Century Collections Online; HCA, High Court of Admiralty; N/A, not available

carried out in London at Execution Dock, and the body then was moved to a suitable place for display in a gibbet. The account of Captain James Lowry's execution in 1752 mentions that his body was conveyed by boat from the scaffold at execution dock to 'The Galleons' (Galleons Reach and Galleons Point are locations on the Thames, north of Woolwich) where he was to be hung in chains.[29] Some accounts record that Lowry's body was later stolen from his gibbet. Earlier in the eighteenth century, the famous pirates Gow and Williams had been executed in 1726 at Execution Dock and their bodies subsequently displayed at Gray's and Blackwall, also locations along the Thames.

By custom, all those sentenced to death by the Admiralty courts were hanged at Execution Dock in Wapping. Following execution, bodies were traditionally chained to a stake at low water until three tides had washed over them, although this practice had apparently died out by the late eighteenth century.[30] Only after this water ritual were those who were eventually to hang in chains taken from Wapping to a location further down the Thames and gibbeted in a place that would be conspicuous to all river traffic entering or leaving London on the east. Those sentenced to dissection were apparently subjected to a scaled-down version of the three-tides punishment and left only until the water touched their toes before being taken to the surgeons, presumably so that the body was not spoiled by time or bloating.

William Clift, of the Royal College of Surgeons, himself painted the scene of an Admiralty execution in 1816. The watercolour shows huge crowds assembled on the Thames foreshore and on boats anchored in the river, watching two figures on a scaffold. Interestingly, the scaffold itself is erected well below the high water mark, and it looks as though it would be possible for the bodies executed there to remain on their scaffold for three high tides without being taken down and restaked in the river. It is also clear that it would make no sense at all to take the bodies down only in order to dip their feet in the water, when a couple of hours of waiting

---

[29] The Monthly Chronologer. *London Magazine*, March 1752, p. 145.

[30] T. Pennant *London; or an abridgement of the celebrated Mr Pennant's description of the British capital and its environs* (London, 1790), p. 157: "The criminals are to this day executed on a temporary gallows, placed at low water mark; but the custom of leaving the body to be overflowed by three tides, has long since been omitted".

would bring the river to their ankles anyway. This accords closely with Pennant's description of the gallows "placed at low water mark".[31]

Smuggling was not a crime that was normally tried by the Admiralty courts; instead, it usually came to ordinary assize courts or the Old Bailey. Dyndor's study of the location of the gibbets of the notorious Hawkhurst gang of smugglers in Sussex and Kent notes that unlike most murderers' gibbets outside London, the men's gibbets were not sited at places of particular significance in relation to the crimes for which they were convicted. Instead, prominence seems to have been a key factor, and gibbets were sited on topographical eminences such as Rook's Hill or Selsey Bill or by main roads. In East Sussex, the gibbets were more likely to be sited close to the villages from which the criminals came.[32]

Liminality: The Symbolic Location of Gibbets

Archaeological studies of unusual burials, such as the deposition of bodies in bogs in northern Europe from the Iron Age to the medieval period, have often suggested that these burials are the remains of criminals whose deviancy is signalled in non-normative burial rites.[33] A key aspect of these interpretations is that the places of disposal of deviant dead are liminal—boglands that are neither wet nor dry; foreshores that are neither sea nor land. Similarly symbolic interpretations of later historical periods are not so common, but there is certainly an argument to be made that gibbeting the criminal body symbolises its liminality and that the enduring nature of the gibbeting process keeps it literally suspended

---

[31] Pennant, *London*, p. 157; Anon. (1761) *London and its environs described* (London: R. and J. Dodsley), p. 289. The copyright holder refused permission to publish this image here.

[32] Zoe Dyndor (2015) 'The Gibbet in the Landscape: locating the criminal corpse in mid-eighteenth-century England', in R. Ward (ed.) *A global History of Execution and the Criminal Corpse* (Basingstoke: Palgrave).

[33] For bog bodies, see R.C. Turner and R. G. Scaife (1995) *Bog Bodies: new discoveries and new perspectives (London:* British Museum Press); P.V. Glob (1965) *The Bog People: Iron-age man preserved.* Trans. Rupert Bruce-Mitford (New York: Barnes and Noble). For liminality in other forms of prehistoric burial, see Liv Nilsson Stutz (2014) 'Mortuary practices' in V. Cummings, P. Jordan and M. Zvelebil (eds.) *The Oxford Handbook of the Archaeology and Anthropology of Hunter-Gatherers* (Oxford: Oxford University Press), pp. 712–28.

between worlds. It is neither buried nor alive; neither human nor thing; and on occasions its landscape positioning also emphasises its liminality. It is not at a place, but by a road. In the case of shoreline gibbets and Admiralty courts, it is at the boundary of land and sea. Whyte and Coolen have both suggested that gibbets occur preferentially at parish or other administrative boundaries.[34]

Technology of the Gibbet

Once a suitable location was identified, the erection of the gibbet scaffold and the suspension of the gibbet cage had to take place swiftly. This was not always easy. The Somerset sheriffs' cravings mention two occasions—in 1739 and 1746—when it was necessary to make holes in the hard rock in order to erect the gibbet pole.[35]

The sheriff was responsible for arranging the erection of a gibbet pole and for the manufacture of a gibbet cage and whatever hooks, chains or other tackle were necessary to suspend the cage. In addition, a pulley or temporary scaffolding would be needed to hoist the heavy iron contraption into position and secure it. In normal provincial practice, gibbets were made for a single criminal and were not normally re-used.[36] Since a gibbeted criminal would be exhibited close to the scene of crime and could remain in his gibbet for many decades, re-use was not normally practical. The sheriff also had to arrange to transport the body from its place of execution to the gibbet site and to organise security if the journey or the process was likely to attract unruly crowds.

---

[34] J. Coolen (2014) 'Places of justice and awe: the topography of gibbets and gallows in medieval and early modern north-western and Central Europe' *World Archaeology* 45(5), pp. 762–79; Whyte 'The deviant dead in the Norfolk landscape'.

[35] Somerset Sheriffs' Cravings (TNA T90/147/307 Stiling; T64/262 Williams and Calway).

[36] However, the gibbet irons in the London Docklands museum, which are not securely provenanced but are likely to come from the riverside area and thus to relate to the Admiralty courts, show two different styles of workmanship. The fact that this cage has apparently been repaired suggests that it might have been re-used. It is possible that re-use was normal for Admiralty gibbets. The other evidence for hanging chains by the Admiralty courts is an image reproduced in Hartshorne, p. 77, showing a very skimpy rig: a simple gusseted chain with a neck brace that would not have secured a body for very long at all. The re-used gibbet and the basic chain are both exceptional designs and might relate to the brief periods of hanging in chains practised by the Admiralty.

Typically, the body of a criminal was gibbeted within a day or two of being executed, but sometimes there were longer intervals, especially when the body had to be transported some distance to the place appointed for gibbeting. Pirates, for example, were usually hanged at execution dock in London but might then be transported many miles around the coast—to Devon or Norfolk, say—to be gibbeted. Occasionally, the judge recognised the time needed to prepare for a gibbeting. Thomas Nicholson, sentenced to execution and hanging in chains at Cumberland Assizes on 22 August 1767, had the date of his execution respited until 31 August in order to make the necessary preparations.[37] Even so, that gave only just over a week to have the gibbet irons made, a gibbet structure created and erected, and a location prepared. Of the 38 cases for which the date of hanging in chains is explicitly stated in the records, 33 were gibbeted on the day of their execution. The other five executions took place between one and four days before gibbeting, and all except one of these five were transported at least 26 miles from the place of execution to the place of gibbeting, so the delay probably is caused by the need to transport the body to the site where the gibbet was erected. Where no separate date for gibbeting is given, as in the majority of cases, it is probable that gibbeting most frequently occurred on the day of execution (Table 2.3).

The Murder Act specifies that capital sentences for murder should be carried out on the second day after conviction. A short interval between sentencing and execution was considered important as a means of increasing the dreadfulness of the punishment and thus its effectiveness as a deterrent and of reducing the occurrence of last-minute pardons, or at least the hope of a last-minute pardon. Henry Fielding believed the great drawback of a long delay between sentencing and execution was that the atrocity of the crime was less raw in the public mind and likely to be overshadowed by the dreadfulness of the punishment.[38] Since the date of conviction is not always known, I have for the purposes of Fig. 2.4 calculated the interval between the first day of the assizes during

---

[37] Assize Calendar Cumberland TNA E389/244/26, 26 August 1767.

[38] G. R. Swanson 1990. 'Henry Fielding and "a certain wooden edifice" called the gallows', in W.B. Thesing (ed.) *Executions and the British experience from the 17th to the 20th century* (Jefferson NC: McFarland and Co), 45–57.

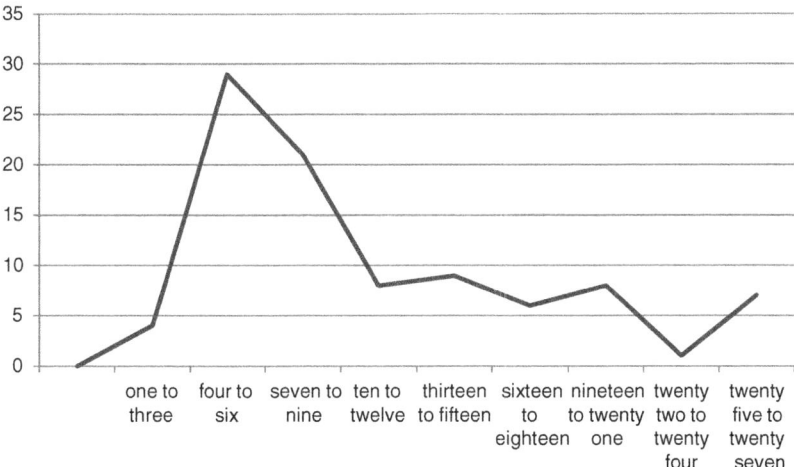

Fig. 2.4 Interval in days between the first day of the assizes during which a criminal was convicted and the date of his execution

which a criminal was convicted and the date of his execution. Among 101 cases in England outside London recorded in the sheriffs' cravings, the mean interval was 10.71 days, although there was considerable variation around this (Fig. 2.4). The Berkshire assizes at which Abraham Tull and William Hawkins were condemned began on 7 March 1787 and they were executed and gibbeted on the 9th—only two days later, or even one day if their case was not heard on the first day of the assize sitting. Thomas Colley, on the other hand, was tried at the Hertfordshire assizes beginning on 29 July 1751 but not executed and gibbeted until 24 August, nearly a month later. Delays of more than two weeks, however, are uncommon. Given that the start of assizes is likely to be before the date of conviction in many cases (assize sittings could take up to a week in this period), we can assume that the blacksmith would normally have a week or less to make a set of irons.

It was necessary therefore to start on the construction of a gibbet and a set of irons as soon as possible after a sentence had been passed. Where possible, the condemned man was measured for his set of irons before execution, a harrowing experience. When Ralph Smith of Lincolnshire was being measured for his irons in 1792, for example, he found it

Table 2.3 Surviving gibbet cages

Date	Name	County	Present location
1720?	Siôn Y Gof	Powys	St Fagan's Museum of Welsh Life
1731	Keal	Lincs	Louth Museum
1742	Breeds	Sussex	Rye Town Hall
Late 18th century?	Anon	London	Museum of London Docklands
1777?	Hill?	Hampshire	Winchester Westgate
1785	Cliffen	Norfolk	Norwich Castle Museum
1786	Matcham	Huntingdonshire	St Ives Museum
1787	Tull or Hawkins	Berkshire	Reading Museum
1791	Miles	Lancs	Warrington Museum
1794	Nicholls	Suffolk	Moyses Hall Museum, Bury St Edmunds
1795	Watson	Norfolk	Norwich Castle Museum
1795	Quin or Culley	Cambs	Wisbech and Fenland Museum
1806	Otter (Temporell)	Lincs	Doddington Hall
1832	Jobling	Northumberland	South Shields Museum (possible replica)
1832	Cook	Leics	Nottingham Galleries of Justice (replica in Leicester Guildhall)

impossible to retain the composure he had exhibited during sentencing, according to a contemporary newspaper report.[39]

Even with the ability to start making the gibbet irons while the condemned man was still alive, there could be considerable time pressure. Moreover, careful measuring of the body to be enclosed was not always possible, and sometimes the condemned man resisted this horrible reminder of his imminent fate. Thus, despite attempts to make the gibbet irons adjustable, designs were not always successful: in 1750, the *London Evening Post* records that the body of John Barchard had to be taken back to the gaol after his execution while the gibbet irons were altered, "they proving too little" (29 September 1750). Some gibbet

---

[39] *Lloyds Evening Post*, 21–23 March 1792, issue 5419.

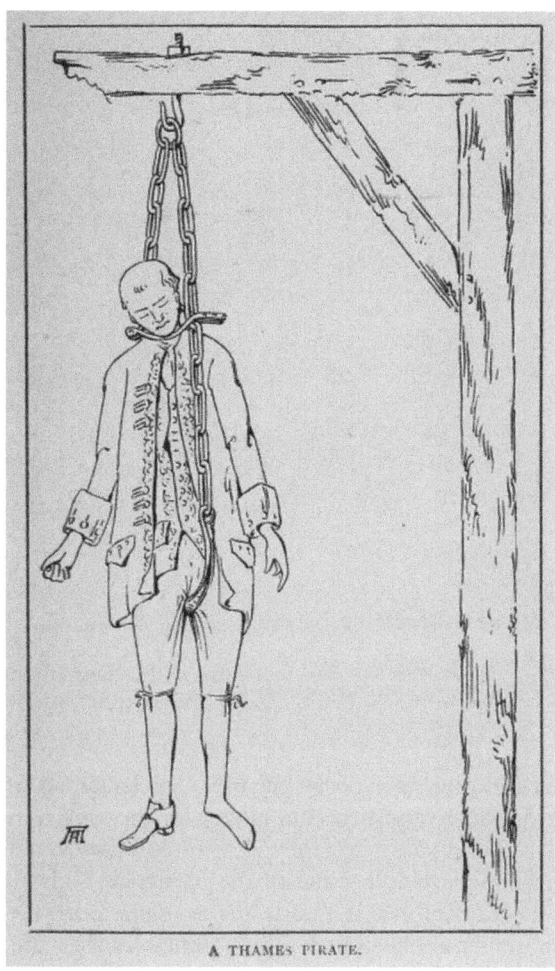

Fig. 2.5 A Thames pirate (from Hartshorne *Hanging in Chains*)

contraptions were so basic that the size or shape of the corpse made little difference. Hartshorne (1891: 77) shows 'a Thames pirate' suspended in what is apparently a single chain with a gusset passing between the legs and a brace around the neck to keep the body upright (Fig. 2.5). It would be easy to remove a body from such a rig, nor would it keep the

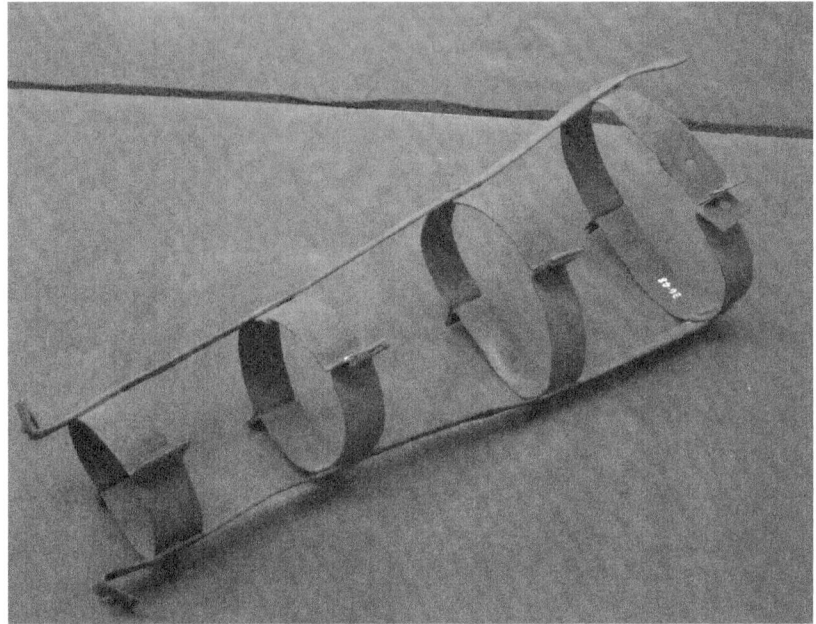

Fig. 2.6 Tull or Hawkins's leg iron, courtesy of Reading Museums (photo: Sarah Tarlow)

body together for long once decay began to accelerate, so such a design could not have been a very successful gibbet for the long term. Surviving gibbet cage structures show that they were often constructed so as to be adjustable to fit the size and shape of the particular body they came to enclose. On the Keal gibbet at Louth, for example, both the belt bands and the long straps are punched several times so that the framework could be extended or contracted and bolted into place to fit supportively close to the body. In the collection of Reading museum, a similar design is evident on the leg iron of Tull or Hawkins's gibbet, which can be tightened to suit the circumference of the leg (Fig. 2.6). Although some gibbet cages, like James Cook's of Leicester, have rigid, hinged hoops, others allow for some degree of shaping to the criminal's body. Only a small part of Gervase Matcham's gibbet survives at Norris House museum, St Ives: part of what is probably a belt, made of a series of five curved and hinged plates which probably would have conformed quite closely to the shape and size of the condemned man's waist.

Usually the fitting of the irons is not described in the sources, but occasionally the cravings mention a cost for having the smith attend the execution in order to fit the irons afterwards. The account of John Curtis's hanging in chains in Wiltshire in 1764 mentions that the smith who made the chains was also responsible for fitting them and was paid to travel to the execution for this purpose.[40] When Rider Haggard discovered the remains of the gibbet and skeleton of Stephen Walton while digging on West Bradenham common, Norfolk, he noticed that the skull had clear scorch marks where it had been burned by a hot iron, thus proving to Rider Haggard that the man must have been dead when enclosed in his gibbet cage and to us that the smith on that occasion fitted the gibbet by soldering or welding.[41] The fact that newspapers commented on the return of John Barchard's body to the gaol after his execution so that the irons could be re-sized suggests that normally the irons were fitted directly after and at the scene of execution.[42] This also constitutes circumstantial evidence that tarring the body was not normally practised, or was a very quick and easy process, since it is hard to see how a corpse could be stripped, immersed in tar, redressed and fitted into irons in a very short time and at the foot of the scaffold.

The poles from which the cages were hung were often very high—10 metres or more, which discouraged attempts to rescue the body or to steal the gibbet—and supported chains which comprised a substantial quantity of iron. The post was also sometimes fitted with spikes around the bottom to make it hard to scale. The gibbet post of Adam Graham, executed in 1748 and hung in chains on Kingmoor, Carlisle, was apparently 12 yards

---

[40] TNA T90/155, Sheriff's Cravings.

[41] H. Rider Haggard (1899) *A Farmer's Year* (London: Longman's, Green and Co), p. 355. Sadly, no scorch marks are evident on the skull fragments which are held today by Norwich Castle Museum, but this may be due to over-zealous cleaning shortly after acquisition.

[42] However, the *Old England Journal* for 28 January 1749 records that the bodies of smugglers Comby, Hammond, Carter and Tapner were returned to the gaol after their execution in order to be hung in chains, which could indicate that the fitting of the gibbet cages was done at the gaol. The body of a fifth man, their partner Jackson, was already at the gaol, where he had died two hours after being sentenced to hang in chains, being "so terribly frightened". It is not possible to say from the newspaper report whether the untimely end of Jackson altered the normal course or location of fitting the irons.

high and had 12,000 nails in it to prevent it being scaled or cut down to remove the body.[43] The sheriffs' cravings for Hampshire in 1761 note that when Francis Arsine was hung in chains the gibbet was "20 feet high made of very strong timber and secured with nails to prevent its being cut down and [fitted with] a secure set of chains". Several London gibbeting accounts (in the sheriffs' cravings) make reference to "plating the gibbet". The fairly detailed accounts for the gibbeting of Thomas Willot in Staffordshire in 1739 include "timber for the gibbet 28 foot long (being 7 yards or thereabout above ground) and cross pieces and carriage there of workmanship of the timber and erecting the gibbet and lining gibbet on each side with bars of iron". Similarly, the cravings account for the gibbet of William Corbett (executed in Surrey in 1764) itemises "the gibbet made strong with iron to prevent it being cut".

The gibbet pole seems almost always to have been made from timber, although sometimes the cravings specify that the timber is "strong" or note that nails or iron bars or plates, as discussed above, should reinforce the main post. The sheriff who commissioned the three-armed gibbet erected for the bodies of Drury, Barker and Lesley in Warwickshire in 1765 lists "materials of stone and timber" for the gibbet, but stone is not usually mentioned in connection with a gibbet, and there is no indication of what its role was to be—perhaps to construct a strong socket for the post. A broken socket stone at Gonerby Hill Foot, Lincolnshire, is believed locally to have supported a gibbet at one time.[44]

Extant Gibbets

I have been able to discover the whereabouts of only 16 extant gibbet cages, despite a thorough literature and online search and an appeal on national radio. The majority of gibbets seem to have disappeared. Those gibbets that do exist are in a variety of styles. The surviving evidence is considered briefly here[45]:

---

[43] Hartshorne *Hanging in Chains*, pp. 66–67.

[44] http://www.lincstothepast.com/photograph/290331.record?pt=S.

[45] See Sarah Tarlow (2014) 'The technology of the gibbet' *International Journal of Historical Archaeology* 18: 668–99 for a fuller discussion. The chains at Weston Park museum, Sheffield, catalogued as Spence Broughton's gibbeting chains are in fact restraining chains and manacles, and another set of chains, perhaps horse furniture. They do not resemble any other gibbet irons and are not included in Table 2.3.

These gibbets show a variety of forms. Some have hooped extensions for arms and legs, others for legs only, and John Keals's gibbet at Louth has only a headpiece and torso, from which the arms and legs would have dangled. Figure 2.7 shows some different styles of gibbet.

The Necessary Functions of a Gibbet

What functions must an effective set of irons fulfil? First, it must contain the body and prevent it from either falling out or being removed, while at the same time still maximising its visibility. In order to do this, most gibbet cages were designed to fit closely to the body, allowing as much as possible of the body to be seen, while ensuring that the gaps between bars were too small to remove it. When possible, the prisoner was measured for his irons before execution, but there were other means of ensuring a close fit, notably construction with punched straps and hoops that could be adjusted to size by riveting (Fig. 2.8). Bodies in advanced decay would necessarily have fallen through the framework in pieces, although the skull, if unbroken, might remain in the headpiece, as in the case of John Breeds at Rye or Sion y Gôf at Dylife (Fig. 2.9). In addition, small pieces of the body could easily be removed by animals or birds. However, by adding to the horror of the gibbeted body, such removals did not diminish the power of the spectacle. In fact, the power of carrion birds around the gibbet to augment the horror was exploited in artistic depictions of the gibbet (Fig. 2.10).

Strength and security seem to have been the most valued and discussed features of a good gibbet. The cravings often describe the gibbet as "strong" and sometimes specify the necessity of making theft of the body impossible. The sheriffs' cravings for Berkshire 1738, for example, mention that the irons of John Sturabout cost £7 and 7 shillings "to prevent [his body] being stolen wherein much iron and workmanship is required". The cravings related to the gibbetings of David Anderson (1736) and William Fairall (1749) in Kent both make explicit reference to the need for security. Anderson's gibbet was "built in a strong manner and filed with nails and braced with iron to prevent the same from being cut down", and Fairall's gibbet was also riveted with iron to prevent his fellow smugglers from cutting

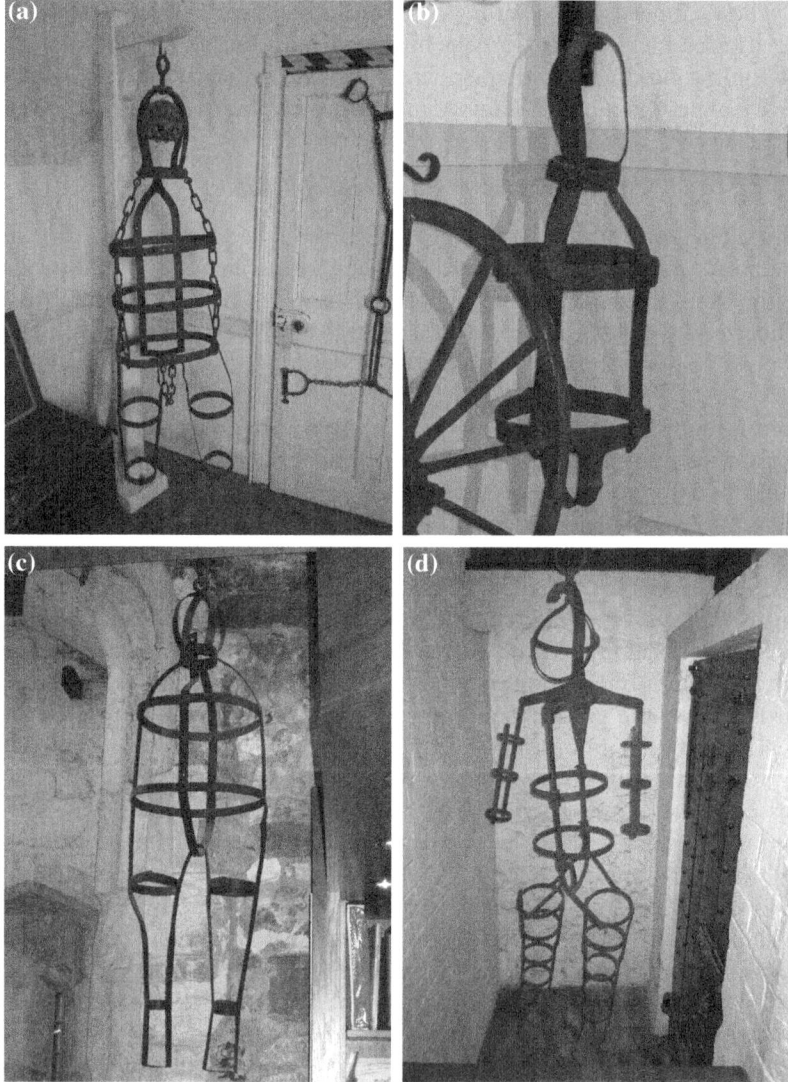

Fig. 2.7 Some different styles of gibbet: a: John Breeds (Rye, 1743, now in Rye town hall); b: John Keal (Louth, 1731, now in Louth Museum); c: possibly 'Jack the Painter' (Portsmouth, 1777, now in Winchester Museum); James Cook (Leicester, 1832, replica now in Leicester Guildhall). All photos: Sarah Tarlow

Fig. 2.8 Multiple punches holes on John Keal's gibbet (photo: Sarah Tarlow)

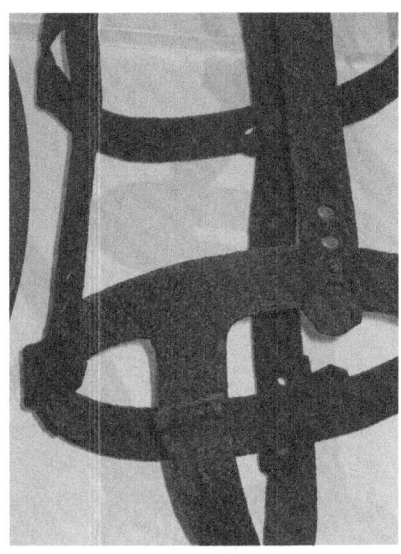

Fig. 2.9 Headpiece of John Breeds's gibbet with large skull fragment remaining (photo: Sarah Tarlow)

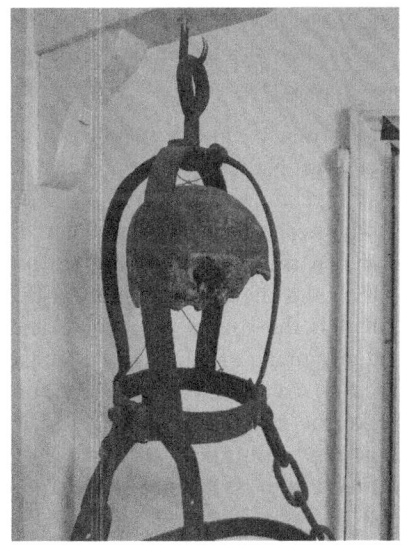

Fig. 2.10 Artistic representation of a gibbet with carrion birds. Vignette from Thomas Bewick's *British Birds* (1804)

him down.[46] The journalist for the *Daily Courant*[47] who reported on John Naden's hanging in chains in Staffordshire was impressed by the chains, made by somebody from Birmingham "in so curious a Manner, that they will keep his Bones together till they turn to Powder, if the Iron will last so long".

Second, the gibbet cage must be conspicuous. This was achieved largely through the use of a tall pole and advantageous siting (sometimes taking advantage of a natural or archaeological eminence such as a hillock or barrow) adjacent to well-used public roads. The successful cage should make the body more visible and more terrible to onlookers. The gibbet cage must contribute to the awe of the spectacle by allowing the body to be seen and permitting some limited movement. If this also caused the

---

[46] TNA T90/147/256 Sturabout; T90/147/118 Anderson; T64/262 Fairall.
[47] *Daily Courant*, 13 September 1731.

chains to creak or clang, so much the better. The Reverend Charles Hardy remembered an occasion in about 1837, driving over West Bradenham Common at night when his horse began to "plunge violently" and he heard a sound above his head: "Clink, clank; clink, clank; clink, clank". His servant told him it was the sound of a gibbet "on which the iron cap and collar of the man, who had been hanged, was swinging to and fro in the October breeze, producing the ghastly sound".[48]

Third, the gibbet cage had to be durable. The body was supposed to remain up there until it had decayed, and as there was no particular time for taking it down, many gibbets remained in their location for decades. Heavy iron was invariably used for the cage. The condition of surviving gibbets is testament to their durability, especially since many of them had been hanging outside for many decades, often followed by a period of burial in wet ground.

Fourth, it had to be possible to construct a gibbet cage quickly. As we have seen, less than a week was available to design and construct the full kit in most cases. The smith had to work together with a carpenter (whose job it was to make the wooden pole), to ensure that the gibbet was securely erected in time for the arrival of the body, and then to encase the body in its irons and probably to oversee its suspension.

Fifth, while being durable and secure, the gibbet cage also had to be light enough to hoist on a gibbet post which could be around ten metres high and not so solid that the visibility of the body was in any way impeded. The criminal on the gibbet should be recognisable to those who had known him in life.

Gibbet Technology and the Absence of Tradition

Gibbeting required a pole/scaffold, a length of chain, and a gibbet cage or suit. The sheriffs' cravings normally bundle all these costs together, but sometimes they specify the recipient of the money and the nature of their job. For example, John Bowland's gibbet, commissioned in

---

[48] Hardy's work, *Social England from eighty years ago to the present jubilee year*, is cited by C.M. in the *Norfolk Chronicle* of 1897 (30 January 1897). This must have been Watson's gibbet of 1795, despite C.M.'s stated belief that a gibbet of such a date would not be standing in the later 1830s. In fact, many gibbets are known to have been still standing fifty years or more after their erection.

Rutland in 1769, cost £5, 15s, 6d for the "set of iron chains", paid to John Fox, a blacksmith, and £6, 2s, 6d for the construction and erection of the wooden gibbet frame, paid to John Wyhters, a carpenter.

The gibbet cage is an unusual artefact. It is comparatively rare—out of only a couple of hundred (at most) that formerly existed in Britain, only a handful are known. The infrequency of its manufacture makes it unusual also. Blacksmiths in Britain during the eighteenth and nineteenth centuries generally made few forms—agricultural implements, craft tools, household objects and farriery (horse-shoeing and horse tack). These artefacts were learned during long apprenticeships and conform to local traditions.[49] By contrast, a gibbet was needed so infrequently that it was not a form within the learned repertoire of most blacksmiths. Moreover, it was needed almost immediately and so left the blacksmith little time to experiment or research other models. Therefore, each blacksmith needed to design a gibbet effectively from scratch. This constant reinvention of the gibbet iron is evident in the proliferation of designs and in the absence of clear typological logic by either region or time period, even though such typologies are observable in other, more frequently made, products of the blacksmith's craft. The range of designs identified represent independent and idiosyncratic responses to the problem of designing a framework which would enable the range of functions identified above.

THE 'CARNIVAL' OF THE GIBBET

Huge crowds are commonly reported in the days immediately following the erection of a gibbet. At least 2000 people are supposed to have visited the scene of Benstead's gibbet on Undley Common, Suffolk on a single day in 1792, and maybe 200 of them made a special ferry crossing in order to visit the site.[50] A similar number was estimated to have attended the gibbeting of Robert and William Drewitt on North Heath Common in 1799; on that occasion, the spectators were accommodated in booths "as

---

[49] J. Bailey (1977) *The Village Blacksmith* (Princes Risborough: Shire); E. J. T. Collins (1996) 'Agricultural hand-tools and the industrial revolution', in N. Harte and R. Quinault (eds.), *Land and Society in Britain, 1700–1914* (Manchester: Manchester University Press) pp. 57–77.

[50] *World*, April 1792, issue 1651.

at a horse race or cricket match".⁵¹ Such numbers, however, were dwarfed by the estimated 40,000 who attended the first day of Spence Broughton's gibbeting near Sheffield or William Smith's on Finchley Common.

What did the gibbet crowd experience? Between the expectation that they would have been unanimously awed and chastened by the spectacle of ignominy and the revisionist position that they subverted the theatre of humiliation, there is a wide spectrum of possible reactions. Gatrell's critique of Laqueur's posited "carnival" of the scaffold questions the idea that the crowd rather than the State was in control of the scaffold experience.⁵² While potentially subversive and "carnivalesque" elements were undoubtedly present at the scene of an execution, the disapproval with which such scenes were described in the contemporary press suggests that it should not be seen as normative behaviour. Gatrell's comments on Laqueur refer only to the scene of execution. How far their debate could be relevant to subsequent hanging in chains will be discussed further in the conclusion. There was no doubt a big difference between visiting a gibbet at or shortly after its erection and encountering a gibbet months or years later. There is no doubt that for many people the gibbet represented simply another destination for a day out. Stephen Monteage of London recorded in his diary for the 16 September 1733: "In the afternoon took a walk with my wife, Mrs Tickling and pretty little Salley to the men in chains upon Stanford Hill".⁵³

If the criminal was sufficiently notorious and interest in his gibbet was great, there was good money to be made from playing to these crowds. The landlord of The Arrow on Clifton Lane next to Attercliffe Common, South Yorkshire, where Spence Broughton was gibbeted, boasted that he had made enough money from the beer sold in the first few days of his exhibition that he was able to retire. A description of the gibbet of William Smith on Finchley Common in 1782 remarks that the 40,000 people who came to view the body the Sunday after his

---

⁵¹ *Evening Mail*, 17–19 April 1799.
⁵² Gatrell *The Hanging Tree*, pp. 90–105, commenting on T.W. Laqueur (1989) 'Crowds, carnivals and the English State in English executions, 1604–1868', in A.L. Beier et al. (eds.) *The First Modern Society: essays in honour of Lawrence Stone*. Cambridge: Cambridge University Press, pp. 305–99.
⁵³ LMA CLC/479/MS00205/001-009.

execution were well fed. Sausages, fried under the gibbet, were available to the more refined class of people who arrived in coaches, chariots and phaetons while the lower ranks, presumably arriving on foot, were sold gin and gingerbread.[54] A similar number apparently attended the Derbyshire gibbet of Anthony Lingard on its first day, and the local vicar, finding nearly all of his parishioners absent from church, decided instead to give his sermon at the site of the gibbet.[55]

A recently erected gibbet seems often to have attracted a carnival crowd which did not always earn the moral approval of the press. The scene at the gibbets of Peter Conoway and Michael Richardson on Bow Common, London in 1770 were widely reported. Several journals disapproved of the erection of drinking booths and the disorderly behaviour of the "mob" at the site. The *General Evening Post* described how "Several people have climbed up the gibbet, and some of them have taken the caps from the malefactors' faces. One fellow had the hardness to call out 'Conoway, you and I have often smoked a pipe together, and so shall we again' on which, to no small diversion to the mob, he climbed up the gibbet with two lighted pipes, one of which he stuck in Conoway's mouth, and the other he smoked as he sat across the gallows".[56] The *Public Advertiser*, meanwhile, opined that the behaviour of the crowd "must give foreigners a shocking idea of the manners of the English" and was appalled that "what is intended as a public example should be treated as a matter of public festivity".[57]

The size of the crowds at the scene of the gibbet fuelled middle-class anxieties about crime and unrest. Robert Hazlitt, hung in chains near Newcastle in 1770 for robbing the mail, expressed shortly before death the desire that his death and display would be "useful to mankind", presumably as a warning against criminal behaviour.[58] However, the proximity of the gibbet seems not to have had a reliably deterrent effect on the criminally minded. In 1826, John Lingard was convicted of assault and robbery committed within sight of the gibbet containing the bones of

---

[54] *Public Advertiser*, 30 April 1782, issue 14927.

[55] C. Drewry (2007) *Wormhill: history of a High Peak village* (Little Longstone: Ashridge Press).

[56] 2–4 August 1770, issue 5744.

[57] *Public Advertiser*, 6 August, issue 11111; 24 July, issue 11106.

[58] John Sykes, *Local Records*, vol. 1.

his brother Anthony, executed eleven years earlier.[59] Nor was this the first serious crime to have been committed at that scene: in 1819, 16-year-old Hannah Bocking chose the road near Lingard's gibbet as the location to give poisoned cakes to Jane Grant, a young woman of her own age who had been offered a job for which Bocking had been turned down.[60] For this crime, Bocking herself was executed and dissected. The same year the *Sheffield Iris* reported a robbery near Attercliffe, pointing out that the "daring offender must have passed through the field in which is the gibbet of the notorious Spence Broughton".[61]

The Curative Power of the Gibbeted Man

Gibbeted bodies not only were magnets for fairs and wild behaviour but also were the unlikely subjects of eighteenth-century medical tourism, sought for their curative and totemic value as sources of healing. The touch of the dead man's hand was believed to cure various diseases and, in former times, had even been recommended by orthodox medical authorities such as William Harvey.[62] There are several accounts of people visiting gibbets specifically to stroke the affected parts of their bodies with the dead man's hand. In 1799, two young women "of genteel appearance" came to the gibbets of Robert and William Drewitt on North Heath in order to have their necks stroked by the hand of one of the dead men in order to cure their scrofula.[63] In the ensuing months, many people travelled to the site of the gibbet with their children in order to hold them up towards the body of Robert Drewitt, who was widely believed to have been wrongly executed, to have his hand passed over their throats. A newspaper search revealed 27 instances of curative uses of the hanged man's hand between 1758 and 1863.[64] It is mostly the newly hanged

---

[59] *The Derby Mercury*, 22 March 1826, issue 4889.

[60] Taylor. *May the Lord have mercy on your soul*, p. 40.

[61] *Sheffield Iris*, 6 April 1818. Thanks to Chris Williams for drawing this report to the attention of the research project.

[62] W. Pagel (1976) *New light on William Harvey* (Basel: S. Karger) p. 50.

[63] *Courier and Evening Gazette*, 24 April 1799, issue 2088.

[64] O. Davies and F. Matteoni (2015) 'A virtue beyond all medicine': The Hanged Man's Hand, Gallows Tradition and Healing in Eighteenth- and Nineteenth-century England. *Social History of Medicine*.

man's hand that was sought after, while still hanging from the scaffold on which the execution had taken place. Touching the hand of a gibbeted criminal must have been challenging, given the height of the post and the rigid design of many cages. The curative use of the bodies of the Drewitt brothers was, in that sense, unusual, and it is possible that the particular draw of those bodies related to the availability of a ladder or a particularly accessible set of irons or both, although such a suggestion is purely speculative. In other ways, however, the gibbet, and the bones contained within it, continued to have power over the bodies, minds and landscapes of the living for many decades, as we shall see in Chap. 3.

Open Access This chapter is licensed under the terms of the Creative Commons Attribution 4.0 International License (http://creativecommons.org/licenses/by/4.0/), which permits use, sharing, adaptation, distribution and reproduction in any medium or format, as long as you give appropriate credit to the original author(s) and the source, provide a link to the Creative Commons license and indicate if changes were made.

The images or other third party material in this chapter are included in the chapter's Creative Commons license, unless indicated otherwise in a credit line to the material. If material is not included in the chapter's Creative Commons license and your intended use is not permitted by statutory regulation or exceeds the permitted use, you will need to obtain permission directly from the copyright holder.

CHAPTER 3

The Afterlife of the Gibbet

Abstract Gibbets could remain standing for many decades. Some were removed because their presence was objectionable; others were eventually brought down by time and the weather. Sometimes, bodies were stolen. Folklore was attached to the locations of gibbets and to the remains which stayed there, and often the names of gibbeted criminals are still attached to places in their landscapes. Parts of the gibbet and of the bodies themselves were collected and curated, sometimes for utilitarian or scientific purposes but often just as curiosities. The case of Eugene Aram's skull is a case in point.

Keywords Afterlives · Folklore · Body parts · Phrenology

How Long Did the Gibbet Remain?

There was no minimum or maximum specified time for a gibbet to remain standing, and they could remain in situ for anything between three days and more than a century. Whereas some were deliberately removed because of the nuisance caused by visitors or because of the offensiveness of the sight and smell of the remains, others stayed in their gibbets until time or weather brought them down. A body that had not been embalmed or otherwise artificially preserved would normally have decayed fully within a few months in the open air, but some bodies became naturally desiccated and survived, entire or in part, for many years. The gibbets of James Price and Thomas Brown, for example, erected on Trafford

Green in 1796, were taken down in 1818, at which time apparently not only nearly all the skeletons remained but also some soft tissue was still surviving.[1] Gibbet cages were normally designed to hold the body quite securely, but as connective tissue decayed, most elements would fall out of the irons. The exception is the skull which was too large to slip between the bars and so is sometimes found still in its position. John Breeds's skull remains inside his gibbet irons, held at Rye town hall. The skull of Edward Corbet, gibbeted on Bierton Common, Buckinghamshire, in 1773 was still visible in his gibbet in 1795 when a correspondent of the *Bucks Herald* noted it during a visit to the Bierton feast. Corbet's gibbet eventually fell when the action of the swivel eroded the attachment and it fell into a ditch.[2]

By the 1830s, the duration of gibbeting had become much shorter—for various reasons. The body of William Jobling, gibbeted in 1832 at Jarrow Slake, near South Shields, was removed without authorisation within three weeks of his execution, supposedly by his relatives or friends, although nobody was ever tried for the offence of removing his body, which, in theory, could result in a sentence of transportation.[3] James Cook, the last man to be gibbeted in England, was executed in Leicester in August 1832, about a week after Jobling. His body was removed only four days after being suspended, following an application to the Secretary of State. In Cook's case, although the correspondence is not published, comment in the newspapers of the time suggests that it was a combination of the huge crowds and the associated possibility of disorder, combined with distaste for the exhibition of cadavers which motivated the removal of the body. The *Leicester and Nottingham Journal* for 18 August 1832 commented,

---

[1] *Lancaster Gazette and General Advertiser*, 2 May 1818, issue 881.

[2] Andrews *Bygone punishments*, pp. 56–57.

[3] *York Herald and General Advertiser*, 8 September 1832, issue 3131 contains the news that his body had been stolen and supposedly buried in the sand. There is more to this than first appears. The *Leicester Chronicle; or, Commercial and Agricultural Advertiser* adumbrated on 11 August 1832, issue 1142, "It is supposed, however, that [Jobling's] fellow workmen will very soon remove [his body] and bury it in some private place ... In the act of parliament ordering murderers' bodies of [sic] to be hung in chains, there is a clause inflicting the punishment of transportation for seven years upon all who may be guilty of stealing the body from the gibbet".

We have heard several reasons given for the interment of Cook's body, but as the Secretary of State's letter has not been published, we can give no positive information on the subject. One cause that we have heard assigned is, that should murders be as frequent within the next twelve years as they have been during the same time gone by, the county would be frightfully studded with such exhibitions, and there being now little waste land except by the side of roads, they must necessarily prove a great annoyance to the inhabitants residing in the villages. However, be the cause what it may, we are glad that the disgusting *sight* has been removed considering it, as we do, the revival of a barbarous custom which a more humanized age has long exploded from the statute book.

WHEN AND WHY DID A GIBBET COME DOWN?

In the absence of any legally specified term for which the body must remain on the gibbet, bodies were generally left until weather, land development or time brought them down. However, there were a number of reasons why a body might be taken down sooner. Local residents sometimes petitioned the sheriff or judge to have a body removed shortly after the gibbeting, and the residents had to give reasons for this. Such reasons divide broadly into two categories: that the gibbet was itself noisome and distasteful, and offended the sensibilities of those who lived or travelled nearby, and that the visitors attracted to the gibbet caused disturbance to the local area.

Concerns of the first kind motivated the complaints about the body of Samuel Hurlock which, in 1747, was removed from its location at Stamford Hill "on Account of the Heat, the Stench of his Body being a Nuisance to the Inhabitants of the Neighbourhood" and placed instead on common land off the Tottenham turnpike.[4] Similar concerns were later made about, for example, Thomas Watkin's gibbet near Windsor (1764) and Jenkin Prothero's near Bristol (1783):

> On Monday last the body of Watkins the Gardener, who was lately executed at Windsor, and hung in chains for the murder of Miss

---

[4] *Old England*, 15 August 1747.

Hammersley's servant maid, was removed from the road side where it hung, and the gibbet erected on the banks of the River, on a complaint that it was a nuisance to the passengers.[5]

Jenkin William Prothero was hanged for murder in 1783 and the judge specified that his body be hung in chains on Durdham Down, Bristol. However, the local inhabitants petitioned the Royal court that the body be moved, and the sheriff of Gloucester was ordered to find a new spot for Prothero's gibbet or to send his body for dissection. The petitioners particularly suggest that the spectacle was revolting to those who sought out the hot wells adjacent to the Down and that the gibbet was "placed so near the back part of the dwelling house of a widow woman who used to let an apartment in the summer season to persons of decent repute from Bristol that it will be injurious to her".[6] The fact that this letter was sent to the sheriff confirms that it was he who normally had responsibility for organising the location of the gibbet. Where the proposed location was on private land, however, the sheriff could proceed only with the permission of the landowner. In the case of the Washwood Heath gibbet, the sheriff omitted this crucial step, and the complaint went directly to the judge.

In 1781, murderers John Hammond and Thomas Pitmore were hung in chains on a shared gibbet on Birmingham's Washwood Heath. The crowds of people attending the gibbeting and visiting the structure afterwards had disturbed a rabbit warren and thus compromised the warrener's livelihood, argued local petitioners, seeking to have the gibbet removed or relocated.[7] As additional argument, the petitioners mentioned the visibility of the gibbet from both Erdington Hall and Aston Hall, illustrating another common factor in the deliberate removal of gibbets: that they offended the sensibilities of polite people. The gibbets of Abraham Tull and William Hawkins in Berkshire were taken down and buried at the request of a local lady. William Andrews recorded that "Mrs. Brocas, of Beaurepaire, then residing at Wokefield Park, gave

---

[5] *St James's Chronicle or the British Evening Post*, 24–26 May 1764, issue 503. His hanging in chains in Gallows Lane near Windsor was reported in the *Public Advertiser* on 13 March 1764. A warrant issued by Judge Wilmot on 30 June 1764 orders the removal of the gibbet and body of Watkins to be hung up again at Churgreen, a mile and a half beyond Windsor towards Reading (TNA E389/243/620).

[6] TNA E389/247/185.

[7] TNA DD/E/208/15, DD/E/208/16, T90/163.

private orders for them to be taken down in the night and buried, which was accordingly done. During her daily drives she passed the gibbeted men and the sight greatly distressed her".[8]

Anthony Lingard's gibbet was taken down by the Duke of Devonshire in response to complaints from local people about the noise the rattling bones (and presumably creaking chains) made.[9] The noise of the gibbet's "creaking cage and bleaching bones" was also noted in relation to an encounter with Spencer's gibbet at Scrooby toll bar, Nottinghamshire, which was erected in 1779 and apparently still visible in 1846.[10]

In 1799, the gibbet of a man called John Haines was controversially sited on Hounslow Heath, occasioning some spirited discussion in the newspapers. The *Whitehall Evening Post* complained that it was situated too close to the road; the *Oracle and Daily Advertiser* agreed that its effect was only "to frighten women and poison travellers"; and the *Morning Post and Advertiser* reported that the royal family were now travelling by a different road to avoid the spectacle. Only the *Morning Herald* demurred, claiming that it was 500 yards from the road and not in sight of any house: a claim made rather suspect by the *Morning Chronicle*'s report that on the night of 16 March the body in its irons was blown from the gibbet into the garden of a nearby house.[11]

THEFT OF BODIES FROM GIBBETS

Despite the possibility of being sentenced to up to seven years' transportation if caught removing a body from its gibbet, friends and relatives of the deceased sometimes attempted rescue. The bodies of Andrew Burnet and Henry Payne, gibbeted at Durdham Down near Bristol, disappeared from their irons a month after their executions in 1744 but were found hidden in some rocks and hung up again. One can only suppose that their rescuers were disturbed or interrupted by the coming of daylight and attempted to

---

[8] William Andrews *Bygone Punishments*, p. 63.

[9] Ebenezer Rhodes (1822) *Peak Scenery*; a letter from Jeffrey Rackett dated 22 March 1826 requesting the gibbet's removal survives in TNA (HO 44/16/25—f25).

[10] Robert Mellors (1920) *Scrooby: The Archbishops' Palace, and the Pilgrim Fathers* (Nottingham: J. and H. Bell).

[11] *Whitehall Evening Post*, 12–14 March 1799, issue 8056; *Oracle and Daily Advertiser*, 26 March 1799, issue 941; *Morning Herald*, 15 March 1799, issue 5769; *Morning Chronicle*, 19 March 1799, issue 9304.

conceal the bodies rather than risk being caught with them.[12] The body of Walter Kidson, also hung in chains in Gloucestershire, on Stourbridge Common, in September 1773, was stolen two years after his execution. *The London Chronicle* of 19 September 1775 (issue 2931) reports that the gibbet was sawn off at the neck and the body removed. Gloucestershire seems to have had an unusual number of gibbet raiders, for it was also in this county that the bodies of Thomas and Henry Dunsden were removed from their gibbets and taken away, on the same night that the lodge of one of the local keepers was raided and a number of deerskins stolen.[13]

In London, in 1759, a body in its irons was stolen from execution dock, where the Admiralty gibbets were located,[14] and a few years later all the gibbets along the Edgware Road were cut down during a single night. This was probably an act of vandalism rather than an attempted rescue, since bodies were left lying in their chains on the road.[15] In 1786, the body of another Admiralty convict—George Coombes, hung in chains at Boar Ness Point, Kent—was stolen, and the Admiralty offered a £50 reward for information leading to the apprehension of those responsible.[16]

In Lincolnshire, the body of Philip Hooton, hung in chains on Surfleet Common in 1769, was stolen about a week after it had gone up, and apparently it was rumoured to have been thrown into the sea. The *Leicester and Nottingham Journal* of 18 March 1769 reported that a reward of £500 had been offered for apprehending those who had stolen the body. Despite the offer of this enormous sum, there is no record of any arrest for this crime. The person who removed the body of John Croxford from Hollowell Heath in Northamptonshire in 1775, nearly eleven years after it was hung up there, was not so lucky. The newspaper records that a man was arrested and prosecuted for the crime.[17] Strangest of all is the case of Gill Smith, hung in chains in 1738 on Kennington Common for the murder of his wife. A week after his execution, somebody cut off one of his legs at the knee and attempted to remove one of his arms, although they were obstructed by his gibbet

---

[12] Darby *Olde Cotswold Punishments*, p. 20.
[13] *Gloucester Journal*, 8 November 1784, issue 3265.
[14] *London Chronicle* 1759, issue 353.
[15] *Lloyd's Evening Post*, 4–6 April 1763, issue 894.
[16] *London Gazette*, 14–18 February 1876, issue 12,726.
[17] *St James's Chronicle or British Evening Post*, 13–16 May 1775, issue 2223.

irons.[18] This is very clearly not an attempt to rescue the body for burial but probably represents the taking of criminal body parts as curios or as a prank.

Weather

For many gibbets, it was neither planned removal nor illegal rescue but the ongoing onslaught of British weather that eventually brought them down. A newspaper correspondent reported meeting a youth in Derby carrying the skull of Matthew Cochlane.[19] Cochlane had been hung in chains fifteen years earlier but his body had finally been blown down by the wind the previous night. "Numbers, who had often stood in melancholy gaze", reported the witness, "repaired to the gibbet, and returned with various parts of his remains". When Tom Otter's gibbet in Lincolnshire was blown down in 1850, 46 years after he was first hung up, the gypsies acted quickly and were able to take nearly all the iron, except for the head piece, which was kept by Edwin Jarvis of Doddington Hall.[20]

More dramatic weather put an end to York's gibbet on Busselton Common near Bristol when lightening split the gibbet "in a thousand little splinters"[21] and allowed the body, which had been hanging for four years, to fall. A gibbet on Hounslow Heath was also struck by lightning in 1768, and one imagines that being tall and prominent structures topped with iron, gibbets were not infrequently struck.

When the body came down shortly after it had been hung in chains, either accidentally or during an attempted rescue, it was sometimes rehung. The body of Captain James Lowry, wrote the *Whitehall Evening Post or London Intelligencer* in 1758, having fallen down soon after hanging, had been brought to Billingsgate where it awaited rehanging. On other occasions, the body would be buried near the gibbet; this is what happened to William Odell, who was reburied "under a gibbet near the hedge on Ealing Common".[22] On nearby Finchley Common, in 1782, Matthew Flood's

---

[18] *Old Common Sense or the Englishman's Journal*, 22 April 1738, issue 64.

[19] *Lloyd's Evening Post*, 28 October 1791, issue 5356.

[20] Jarvis recorded the event in a commonplace book which is still kept at the hall in the possession of Jarvis's descendant Claire Birch.

[21] *London Evening Post*, 29 June–2 July 1745, issue 2754.

[22] *Public Advertiser*, 10 January 1761, issue 8170.

gibbet, which had been erected sixty years earlier, was clandestinely sawn down and left near the remaining stump of gibbet post, after two of his fingers had been removed.[23]

Enclosure and Convenience

Since many gibbets were situated on common land, the enclosure of the commons, which was proceeding swiftly in much on England and Wales through the later part of the eighteenth century, precipitated the removal of gibbets. This is what happened at Badley Moor, Norfolk, for example, when James Cliffen's gibbet was removed as part of the enclosure process. Whyte notes that the gibbets of Stephen Watson on West Bradenham Common and William Suffolk on North Walsham Common, as well as Cliffen's, were taken down in the same year that their parishes were enclosed.[24]

GIBBET LORE

A quantity of local lore exists around gibbets and some stories recur in several guises. One common motif is the bird nesting in the human remains. *Machie's Norfolk Annal*, vol. 1, 1800–1850 records that around 8 June 1801 a starling's nest with young birds in it was taken "out of the breast of Stephen Watson, who hangs on a gibbet on Bradenham Common, near Swaffam" (p. 6); another starling's nest was found in the chest cavity of Gabriel Tomkins at Dunstable,[25] and the baby birds were removed and sold as curiosities by a man who broke one of Thompson's ribs to get at the chicks. In the skull of James Price, gibbeted on Trafford Green, Cheshire, in 1796 was found the nest of a wren or a robin.[26] An unspecified bird was said to have nested in the skull of John Stretton, whose gibbet on Finchley Common was blown down in 1776.[27] A commonplace book kept by Edwin Jarvis of Doddington Hall records how a "willow-biter" (blue tit) made its nest in the mouth of the body of Tom

---

[23] No explanation is given for this curious incident, which was reported in the *London Chronicle*, 4–6 June 1782, issue 3981.

[24] Whyte "The deviant dead", p. 25, 37.

[25] *St James's Chronicle or the British Evening Post*, 22–24 June 1762, issue 201.

[26] A wren, according to the *Lancaster Gazette and General Advertiser* for 2 May 1818 (issue 881), or of a robin, according to www.mickletrafford.org.uk/history.php.

[27] *St James's Chronicle or the British Evening Post*, 21–24 December 1776, issue 2463.

Otter (executed in 1806) about a year after he was hung up (Fig. 3.1). A similar story relates to Bennington in Norfolk. Jarvis records the riddle made about the nest in Tom Otter's skull:

> There were nine tongues all in one head
> The tenth went out to get some bread
> To feed the living in the dead.

One of the most entertaining pieces of gibbet lore, and one that demonstrates the general aversion to gibbet sites, especially during the night, is the widespread story of the talking gibbet. This folk story typically features a man bragging of his courage to his fellows at an inn. He then volunteers or is dared by the landlord or his companions to visit a nearby gibbet in the middle of the night and greet the body hanging there and perhaps also to offer the criminal hanging there some food or drink. As he goes to carry out his task, the braggart feels his courage begin to fail but steels himself to offer some soup or ale to the grisly remains. But he is terrified when the body in the gibbet answers him back, and immediately runs away. The usual twist is that the voice of the dead man was actually that of one of his drinking companions who had rushed to the gibbet ahead of him and hidden himself nearby. Such tales attach to the gibbet of John Grindrod (executed 1759) on Pendleton Moor, Lancashire; Matthew Cocklane, executed in Derby in 1776; and others.[28] There are persistent stories of criminals gibbeted alive during this period, but none of them can be substantiated. The case of John Whitfield, a highwayman gibbeted in Cumbria in 1769, for example, is cited by Andrews as a case of gibbeting alive.[29] However, contemporary accounts, such as that in the *St James's Chronicle* for 12 August 1768, record that Whitfield was executed at Carlisle before being hung in chains near Armithwaite. Gibbeting alive was still practised in the eighteenth century in the Caribbean and parts of America as a punishment of

---

[28] Andrews, *Bygone Punishments*, pp. 51–52. It is possible that Grindrod's story is the original because it was the subject of a popular ballad that was published in 1855 in W. Harrison Ainsworth's *Ballads: Romantic, Fantastical and Humorous*, and it is certainly plausible that variants of this pleasing story were attached to gibbets in other localities.

[29] Andrews, *Bygone Punishments*, p. 58.

Fig. 3.1 'Willow biter' and rhyme, drawn and recorded in the commonplace book of Edwin Jarvis of Doddington Hall, Lincs., courtesy of Claire Birch (photo: Sarah Tarlow)

slaves for crimes or acts of rebellion but is not known for Britain during this period.[30]

THE MATERIAL AFTERLIVES OF THE GIBBET

The material remains of the gibbet, including the wooden framework, the iron work and the human bones, followed various journeys in their own afterlives. The wooden gibbet post and cross element were often

---

[30] Gibbeting alive seems to have been practised in Britain during the sixteenth century. William Harrison's *Description of Elizabethan England* (1577) notes that most felons sentenced to death are cut down and buried when they are dead, "But if he be convicted of wilful murder, done either upon pretended malice or in any notable robbery, he is either hanged alive in chains near the place where the fact was committed (or else upon compassion taken, first strangled with a rope) and so continueth till his bones consume to nothing" (Book III, Chap. 6).

substantial pieces of wood, as we have seen in Chap. 2, and could be ten metres or more in length. After functioning as gibbet posts for several years, they were sources of well-seasoned large timbers which were desirable for many utilitarian purposes. The post that had served to suspend Eugene Aram's gibbet in Knaresborough was installed in a nearby inn, the Brewer's Arms, formerly known as the Windmill, where it served as a beam.[31]

The wooden posts were also of interest because of their former grisly function. An 1867 report in the *Times* notes the interest generated by the rediscovery of the post of Spence Broughton's gibbet in Sheffield:

> Discovery of Spence Broughton's Gibbet
>
> The remains of the Gibbet-post of Spence Broughton, who was hung in irons on Attercliffe Common after being executed at York for the robbery of the Sheffield and Rotherham Postman, have this week been dug out of the ground.
>
> It is solid old oak, perfectly black and quite sound, though embedded in the ground since 1792. It consists of a massive framework, 10ft. long and 1ft. deep, firmly embedded in the ground to support the Gibbet-post, which passed through it's centre and was bolted to it. Some 4ft. 6in. of this post is left, the remainder of the post is 18in square.
>
> This relic was discovered by a person named Holroyd, in making excavations for the cellars of some houses in Clifton Street, Attercliffe Common, opposite the "Red Lion". It was conveyed into the garden of that Inn, where it may now be seen.
>
> Many hundreds of persons have paid it a visit.[32]

The current location of the post is not known, but the association with Spence Broughton's gibbet has been re-invented in the present-day Noose and Gibbet pub on Broughton Lane, Sheffield, which is decorated with a (highly fanciful) gibbet (Fig. 3.2).

---

[31] P. Walker (1991) *Murders and Mysteries from the Yorkshire Dales* (London: Robert Hale), p. 83. According to the trade directories, there has been no Brewer's Arms in Knaresborough since the 1910 s and I have been unable to find its exact location.

[32] *The Times*, 6 May 1867.

Fig. 3.2 'Noose and Gibbet' pub, Sheffield (photo: Tom Maskill)

The remains of the gibbet post that had been used for Andrew Mills, hung in chains for murder in the later seventeenth century in County Durham, was known as "Andrew Mills's stob". Pieces of the stob were taken away as charms for curing toothache, until there was nothing left.[33] Ralph Smith's gibbet post, erected in Lincolnshire in 1792, was used to make various fancy goods, including a tobacco bowl, now in the Guildhall museum at Boston.[34]

It is likely that the ironwork of gibbets was frequently recycled for its value as scrap metal, as was presumably the case with the irons of Tom Otter's gibbet cage, which were taken (by "gypsies", according to a local source) very soon after the gibbet was blown down. It is possible that gibbet iron was recycled into items that gained part of their value from

---

[33] Andrews, *Bygone Punishments*, p. 47.
[34] www.boston.gov.uk/index.aspx?articleid=4138.

their glamorous association with criminal notoriety. Anthony Lingard's gibbet irons, for example, were made into toasting forks.[35]

The bones of gibbeted criminals usually did not survive but were broken, dispersed by animals and decayed by natural processes. Archaeologically, a couple of possible gibbeting deposits are known, marking the probable locations of gibbets. Disarticulated bone was found during the enlargement of the Royal Edward Dock at Avonmouth, Gloucestershire, at the beginning of the twentieth century. The bones are thought to have originated from the gibbet that stood nearby on Dunball Island, possibly that of Matthew Mahoney, executed in 1741, which was blown down in a storm in the late 1830s. Similar remains from Eyre Square, Galway City relate to a place of gibbeting. In Ireland, the display of criminal bodies and body parts in urban locations was more frequent than in England, where gibbets were almost always erected in the countryside.[36] The paucity of post-gibbeting or post-dissection human remains traceable by either historical or archaeological sources is itself interesting. It evidences the successful disintegration or obliteration of the criminal body.

Some remains are known to have been buried after the gibbet fell or was removed, most usually in a pit at or near the gibbet site.

The body of John Gatward, gibbeted probably at Collier's End near Puckeridge in Hertfordshire, was eventually buried by his mother, according to one source:

> I saw him hanging in a scarlet coat, and after he had hung about two or three months it is supposed that the screw was filed which supported him and that he fell in the first high wind after. Mr Lord of Trinity passed by as he lay on the ground, and, trying to open his breast to see what state his body was in, not being offensive but quite dry, a button of brass came off, which he preserves to this day... His mother had the body brought to the inn and buried it in the cellar.[37]

---

[35] Andrews, *Bygone Punishments*, p. 71.

[36] J. Brett (1996) Archaeology and the construction of the Royal Edward Dock, Avonmouth 1902–1908. *Archaeology in the Severn Estuary* 7: 115–20. C. Lofqvist (2004) *Osteological report on human skeletal remains from Eyre Square, Galway City* (Moore Archaeological and Environmental Services Ltd, unpublished report).

[37] Cole's manuscript history of Cambridgeshire, cited in Charles George Harper (1908) *Half-hours with the Highwaymen: picturesque biographies and traditions of the knights of the road, volume 1* (London: Chapman and Hall), pp. 202–04. *The London Magazine or Gentleman's Monthly Intelligencer* (vol. 26, p. 202) records that at Hertford Assizes one John Gatward alias Gardgreen was convicted of robbing the Northern Mail. He was sentenced to be hung in chains at New Bridge, Puckeridge, Colliers End.

On other occasions, bones were kept either as gruesome but thrilling curios or for their phrenological interest. The alleged skull of Michael Morey, gibbeted on the Isle of Wight in 1737 for the murder of his grandson, was an attraction to guests at the nearby Hare and Hounds tavern until recent archaeological examination confirmed that the skull in question is female and probably belonged to one of the individuals buried in the Bronze Age "Michael Morey's tump" on which Morey's gibbet was situated. Morey's gibbet post is still incorporated into the fabric of the inn, and a notice adjacent to the beam gives its provenance.

Four men were hung in chains near Guyhirn in the Fens near Wisbech in 1795. Their gibbets were eventually washed down by a flood coming in from the Wash in 1831. A local diarist recorded that his brother, Joseph Peck of Bevis Hall, acquired the headpiece of one of the gibbet cages.[38]

As this brief review shows, the material remains of the gibbet were conserved and re-used not only for their utilitarian value as building elements or scrap metal but also for their glamorous association with the body of the criminal. The body itself, unless it was salvaged by friends or family and buried, had value as a curio or scientific value as an object of phrenological investigation. A closer look at the material afterlives of one eighteenth-century celebrity criminal will demonstrate the complex and multiple ways that the power of the criminal body—and his gibbet—endured after death.

Bodies and Body Parts: Eugene Aram

Eugene Aram, hung in chains at Knaresborough in 1759, was not a typical eighteenth-century murderer. He was an educated professional, a published author of works of philology who, at the time of his arrest at a school in King's Lynn, was working on his comparative lexicon of Latin, Greek and Celtic.[39] Eugene Aram was born in 1704 to a family of

---

[38] Diary of John Peck 1818, p. 134, held by Wisbeck and Fenland Museum, which also holds the gibbet headpiece.

[39] This history of Eugene Aram is largely compiled from 'The genuine account of the life and trial of Eugene Aram', *The Critical Review, or, Annals of Literature* (September 1759) 8: 229–238; N. Scatcherd (1838) *Memoirs of the celebrated Eugene Aram*, (London: Simpkin, Marshall and Co.); J. Dobson (1952) 'The College criminals 2: Eugene Aram', *Annals of the Royal College of Surgeons of England* (April 1952) 10(4): 267–275.

labourers in Yorkshire in the north of England. His unusual intellectual energy and quick mind enabled him to gain an education and to discover and develop a particular gift for languages, especially ancient ones.

Aram was hanged for the murder of Daniel Clark, a shoemaker, who had disappeared in 1745. When Aram precipitously left Knaresborough, not long after Clark had vanished, his friends assumed that he had fled with a quantity of valuable goods he had acquired illegally. Thirteen years later, the discovery of bones in a cave just outside Knaresborough led to speculation that Aram and another man, Richard Houseman, had conspired to kill Clark and steal his possessions. Aram was traced and arrested; Houseman, who in all accounts seems far more suspicious, turned King's Evidence and testified that Aram had murdered Clark. At his trial, in August 1759, Aram decided, unwisely, to conduct his own defence.[40] He questioned the identification of the bones and asserted his own good character but did not challenge the shaky, inconsistent and unreliable evidence of Houseman, his former friend. Despite the lack of conclusive evidence, he was convicted and sentenced to death and to have his body hung in chains close to the scene of the murder, on the wooded banks of the Nidd gorge at Knaresborough. According to criminal defence attorney Rodney Noon, it is very unlikely that any contemporary court would convict on such evidence or that such a conviction would be safe enough to withstand an appeal.

Accordingly, Eugene Aram was executed at York, after an unsuccessful attempt to end his own life in prison, and his body returned to Knaresborough, where his gibbet was erected close to the scene of crime, overlooking the river Nidd; his body remained there, gradually decomposing, for at least 25–30 years.

Aram's crime and trial were of great public interest. The association between the apparently gentle and scholarly man and violent murder for material gain was unusual and—combined with the instability of the evidence on which he was convicted—resulted in a widespread belief that the wrong man had been executed. His biographer Norrison Scatcherd

---

[40]Rodney Noon, 'Should Eugene Aram have Hanged?' *Web Mystery Magazine* 1(1), Summer 2003. http://lifeloom.com/Eugene_Aram.htm. Accessed 25 March 2013.

described the riots and threats with which Houseman was greeted on his return to Knaresborough. Aram's story was irresistible to cultural producers of the period. Bulwer-Lytton's novel *Eugene Aram* (1831), giving Aram a beautiful and brilliant lover, romanticised the story. Bulwer-Lytton's Eugene Aram, though involved in the death of Clark, was the victim of circumstances and no murderer. The novel was adapted for the stage and had a successful run with Henry Irving in the title role. Thomas Hood's narrative poem "The Dream of Eugene Aram" (1829) was recited by generations of schoolchildren. PG Wodehouse even has Bertie Wooster quoting Hood's poem in proper Wooster style:

> All I can recall of the actual poetry is the bit that goes: Tum-tum, tum-tum, tum-tumty-tum, I slew him, tum-tum tum! (PG Wodehouse, *Jeeves Takes Charge*, 1916)

Hood's Aram, though guilty, was thoughtful, penitent and intelligent: a sympathetic hero. Bulwer-Lytton's novel and Hood's poem are the best known of Aram's literary incarnations, but there are many more,[41] including a stage play and at least three films.[42]

Eugene Aram's body remained in the gibbet for some years. One account holds that his wife collected his bones as they fell from the gibbet; if true, this account suggests quite a turn-around in her feelings about her late husband, who had abandoned her and in whose arrest she had been instrumental.

At some point, probably before the end of the eighteenth century, a doctor called Hutchinson, then practising in Knaresborough, decided to augment his private cabinet of curiosities with the skull of Eugene Aram and managed to remove it from its gibbet cage.[43] Writing in 1832, the pseudonymous correspondent of a literary journal imagines Hutchinson's attempt to extract the skull:

---

[41] See Judith Flanders (2011) *The Invention of Murder: how the Victorians revelled in death and detection and created modern crime.* Harper Collins; Nacy Jane Tyson *Eugene Aram: the literary history and typology of the scholar criminal* (1983).

[42] The play, by W.G. Wills, opened in 1873 with Henry Irving in the lead role; the films were by Edwin Collins (1914), Richard Ridgeley (1915) and Arthur Rooke (1924).

[43] Norrison Scatcherd (1838) *Memoirs of the celebrated Eugene Aram who was executed for the murder of Daniel Clark, in 1759: with some account of his family and other particulars* (London: Simpkin, Marshall and Co.).

on a dark and stormy night, agitated by conflicting feelings, like a bridegroom on the eve of marriage, the doctor sallied forth, from the town of Knaresborough, with a ladder on his shoulder, and with the firm purpose of mounting the gibbet and detaching from the iron hoop which bound it the skull of Eugene Aram. The gibbet clung to its own property with wonderful tenacity; but the ardor of the doctor became a furor, and he succeeded in extricating another neck, almost at risk of his own.[44]

Why was Hutchinson was so keen to acquire Aram's skull? It is probable that he wanted it simply as a curiosity because of its association with a significant local event—and one which had attracted national attention. However, it is as evidence for the new "science" of phrenology that Aram's skull became best known. If the correspondent of the *Phrenological Journal* of 1839 is right that Scatcherd had seen the skull in Hutchinson's possession forty years earlier, then it is unlikely that phrenological study was a motivation for its original acquisition, as phrenology became popular only following the publications of Gall and Spurzheimer in the early nineteenth century. Indeed, Simpson claims that Hutchinson was only "desirous of possessing the skull of so noted a person as Eugene Aram" (1839: 67). However, within a few decades, the skull was important not only as a phrenological specimen but also as a test case on the interpretation of which turned the credibility of phrenology as a whole.

The skull resided in Hutchinson's personal museum until he died. On Hutchinson's death, the skull passed to his widow's second husband, and his former assistant, Mr Richardson, a surgeon from Harrogate. When, in 1837, the young Dr James Inglis, burning with phrenological zeal, took up a post as physician at the public dispensary in neighbouring Ripon, it is probable that he found out about Aram's skull from Richardson, as a fellow medical man working in a neighbouring town. It was Inglis who presented the skull to the Newcastle meeting of the British Association for the Advancement of Science in 1838.

Phrenology divided the British scientific establishment. Some strong voices maintained that such hokum had no place among the Fellows of the British Association for the Advancement of Science; others, equally strong, saw it as a progressive and rigorous approach to understanding character and the workings of the brain. The skull passed from Dr Richardson to his step-grandson, John Walker, in whose private

---

[44] 'Civis' (1832) *The Literary Gazette*, 14 January 1832, p. 25.

collection it remained, first at Malton in Yorkshire and then at Great Yarmouth, Norfolk, when Walker moved house. He presented the skull to the Royal College of Surgeons (Dobson 1952) in 1869, by which date it had become something of a strange embarrassment to its owner, an Anglican minister, who therefore sought to place it in a museum. The skull was included in Sir William Flower's catalogues of the Royal College collections in 1879[45] and 1907. The skull remained in the museum of the Royal College until 1993 when it was given to King's Lynn Borough Council, which exhibited it in the Old Gaol House museum in the town of Lynn, where it remains on public view at the time of writing (Fig. 3.3).

Phrenology at the 1838 British Association Meeting

When Inglis presented Aram's skull to the medical delegates at the British Association meeting in Newcastle in 1838, phrenology was not

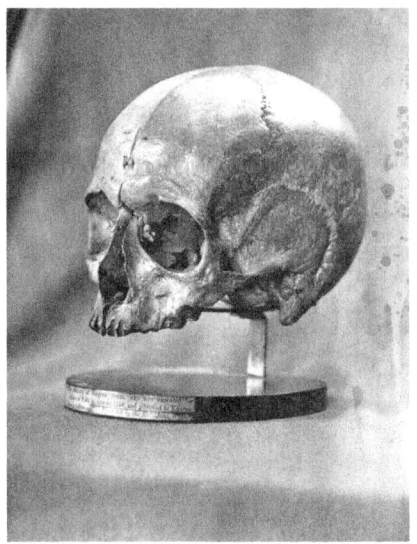

Fig. 3.3 Eugene Aram's skull (photograph courtesy of King's Lynn Museum)

---

[45] *Catalogue of the Specimens illustrating the Osteology and Dentition of Vertebrated Animals*, p. 49, entry 337.

universally accepted as a science, and indeed it was always treated with suspicion and scepticism by many, or indeed most, of the British scientific establishment. Accounts of the 1838 meeting are mostly unsympathetic. This one, for example, is from the *Literary Gazette*:

> the Doctors had a dose of phrenology foisted into their section; and hardly has that science made a more absurd appearance since Tony Lumpkin practised it upon Crackskull Common.

James Inglis trained in Edinburgh, which was a stronghold of phrenology in Britain; the Edinburgh Phrenological Society, established in 1820, was the first such society in Britain. Phrenological societies were established in Wakefield (1825) and Manchester (1830) as well as London (1823), Birmingham (1838) and Aberdeen (1838), so there was some regional support for Inglis's position. But the 1838 meeting of the British Association for the Advancement of Science was also a significant moment in the history of phrenology in this country. Because the B.A. had excluded phrenology from the disciplines it recognised as properly scientific, the newly formed Phrenological Association held its own parallel meetings in the same city (Newcastle) and at the same time. Therefore, the presentation and analysis of Eugene Aram's skull were crucial in negotiating the respectability of the science. Van Wyhe has noted that phrenologists depended very heavily on single examples to legitimate their approach, rather than employing any kind of quantitative or experimental method. Aram was an ideal example, and phrenological discussion of his skull an entirely circular exercise. Since it was precisely his character that was in dispute (gentle scholar or hardened murderer?), any phrenological interpretation could be fitted to the story. Analysis of his skull did nothing to prove or disprove the method.

Aram was a celebrity criminal. Although he was convicted of a murder whose motivation appeared to have been purely monetary, his life and character did not fit the normal stereotype of a violent criminal. He had not lived the life of a thug but that of a scholar, a teacher and a man of apparently refined sensibilities, all of which both interested the public and occasioned later doubts about his guilt. Fictionalised retellings of his life, crime, flight from justice and eventual trial and execution took different positions on Aram's culpability, but all portrayed him as an intelligent and reflective character (Fig. 3.4). Aram's conviction would certainly not be regarded as safe today and even at the time his guilt was widely doubted. Given the fame and popularity of his legend, there was great

Fig. 3.4 Gustave Doré's engraving of Eugene Aram (photostock)

public interest in the truth of the Aram story, and scientific examination of his skull therefore afforded a method by which the question of his likely criminality could be addressed. Was he "a criminal type"? Aram's fame was not the only kind of "afterlife" he enjoyed; his actual body continued to be a thing of powerful and changing meanings long after his final breath.

Open Access This chapter is licensed under the terms of the Creative Commons Attribution 4.0 International License (http://creativecommons.org/licenses/by/4.0/), which permits use, sharing, adaptation, distribution and reproduction in any medium or format, as long as you give appropriate credit to the original author(s) and the source, provide a link to the Creative Commons license and indicate if changes were made.

The images or other third party material in this chapter are included in the chapter's Creative Commons license, unless indicated otherwise in a credit line to the material. If material is not included in the chapter's Creative Commons license and your intended use is not permitted by statutory regulation or exceeds the permitted use, you will need to obtain permission directly from the copyright holder.

CHAPTER 4

Conclusions: Why Gibbet Anyone?

Abstract Given the very high cost of hanging somebody in chains, why was it ever carried out? It was intended to make a deterrent impression on potential criminals and to demonstrate the power and order of the State. However, the many and variable responses to hanging in chains meant that the practice did not always have the intended effect. Gibbetings were infrequent and memorable and served to make the names and histories of those so treated memorable and enduring. Even the very last occasions of hanging in chains were massively popular events, so the distaste expressed by some newspaper commentators was not universally shared.

Keywords Costs of gibbeting · Body · Punishment · State · Power

The Costs of Gibbeting

Gibbeting was an expensive business. We have very good knowledge of the costs from the sheriffs' cravings. Where it is possible to disaggregate the costs of gibbeting from the overall costs of execution, the mean cost per gibbeting of 71 costed cases dating between 1736 and 1799 is £16, with a range from £2 10s to £56 12s. There was a considerable asymmetry in costs between gibbeting and dissection. In the Midlands Assizes Court Circuit covering Derbyshire, Leicestershire, Lincolnshire, Northamptonshire, Rutland and Warwickshire (including those hanged at Coventry, Derby, Lincoln, Leicester and Nottingham), the average costs reclaimed by the sheriff for organising a basic dissection was £5 11s by the 1770s. Moreover, in the case of dissection, the sheriff could sell the condemned for a supply fee (of around £5), still reclaim centrally the basic cost of organising the dissection (£5 11s), and be actually in profit.[1]

Since the surgeons could make money from staging different types of dissections over a period of up to four days after getting the body, they did not necessarily haggle too much about any supply fees. Surgeons could make as much as £80 in notorious homicide cases from audience entrance fees; in a less renowned murder case, about £40 seems to have been average.

Thus, there is a great discrepancy between dissection—potentially quite a lucrative event for both sheriff and surgeon—and gibbeting, which almost invariably meant that the sheriff was out of pocket, since expenses were rarely repaid in full.

Given the expense, the distastefulness and the practical difficulties of gibbeting, in addition to the generally high demand for healthy young bodies supplied at predictable times to the dissection rooms, lecture halls and private parlours of Britain's medical men, why was the spectacle of hanging in chains enacted at all? And why did such a barbaric practice have its period of greatest use and legal enshrinement during the "Age of Reason", in contrast to any hypothetical "civilising process" or triumph of neat self-discipline? How and why did the practice finally come to an end? Finally, what does the historical practice of hanging in chains tell us about attitudes to bodies, to the dead, and to criminals in the eighteenth and nineteenth centuries?

The Murder Act: An Anachronism?

The Murder Act, notes Cockburn, seems to encapsulate the inconsistent, incoherent and contradictory attitudes towards bodily punishment in the mid-eighteenth century. It nods to both reformist and traditional philosophies of punishment: it extends and enshrines the use of brutal and public punishment at the same time that, for example, punitive whipping was beginning to be taken out of the public arena and carried out in

---

[1] I am indebted to Elizabeth Hurren for the data on the economics of dissection.

private.[2] Moreover, the Murder Act formalised in law the well-established practice of hanging in chains, a spectacular and enduring post-mortem punishment, at a moment when such punishment had already passed its peak frequency; gibbeting in Britain would never again reach the levels of the 1740s—the decade before the Murder Act came in.

According to the traditional narrative of punishment, by the mid-eighteenth century, spectacular and horrific treatments of the criminal body had largely given way to more private and humane bodily punishments. Foucault, for example, sees the replacement of blood-thirsty punishment by reformatory discipline as a form of social control.[3] By contrast, Spierenberg, while leaving the essential chronological trend intact, thinks that rather than the working of social power, this transformation relates more to a cultural shift of sensibilities through which a new kind of affective and empathetic individualism produced an aversion to public suffering.[4] The Murder Act was passed during, and was a significant part of, the mid-eighteenth-century crisis in attitudes towards execution. According to McGowen, although there was no major change in practice, perceptions of the event(s) of execution rapidly became more complicated and ambivalent. After the 1750s, says McGowen, "the gallows regime was less securely anchored". The public exhibition of the body, like the public execution, provoked a range of responses, including both an acknowledgement of State power, or of social justice done, and unease or revulsion in the face of human suffering or macabre spectacle.

Steven Wilf interprets the privatisation of punishment as the outcome of an *aesthetic* preference not to lessen the horror of suffering but to achieve a salutary effect through cultivating the imagination of the crowd rather than stimulating their senses. He dates this process to the 1770s and '80s, following the failure of an attempt in the 1750s to renew and revivify the spectacle of public punishment.[5] In this view, the aim of the Murder Act

---

[2] J.S. Cockburn (1994) 'Punishment and Brutalization in the English Enlightenment', *Law and History Review* 12(1): 155–79, 171–72.

[3] Michel Foucault (1974) *Surveiller et punir: naissance de la prison* (Paris: Editions Gallimard). See also Michael Ignatieff (1978), *A just measure of pain: the penitentiary in the industrial revolution 1750–1850* (New York: Pantheon Books) for a similar argument.

[4] Pieter Spierenberg (2008) *Executions and the evolution of repression from a preindustrial metropolis to the European experience* (Cambridge: Cambridge University Press).

[5] Stephen Wilf (1993) 'Imagining Justice: aesthetics and public executions in late eighteenth-century England', *Yale Journal of Law and the Humanities* 5(1): 51–78.

was "to heighten the terrifying aspect of execution aesthetics". The case of hanging in chains, however, would seem to stimulate both the imagination and the senses of the crowd, as will be discussed below.

The historiography of the body in the long eighteenth century has not dealt directly with the practice of hanging in chains. Cultural historians of the body have concentrated instead on spectacular bodily punishments, including executions devised to maximise bodily pain. The punishments studied by Foucault and others are vengeful, brutal acts carried out on a living body—at least a body that was living at the start of the process. A second tradition of historical interpretation revolves around the practice of anatomical dissection: here the body is dead but it is examined, mapped, known through a nexus of power relations, ritualised performances and scientific curiosity. Hanging in chains does not fit into either tradition of bodily punishment, although it partakes of both.

The practice of hanging in chains, then, might have been intended to accomplish several things:

1. According to the Murder Act, to function as a sufficient "Mark of Infamy" to deter the crime of murder.
2. To make a vivid and salutary impression on the masses.
3. To act as a collective act of restitution and restoration of order.
4. To cement a memory and become part of communal historical knowledge.
5. To demonstrate State power.

The degree to which the gibbet successfully fulfilled any of these functions is unclear but will be considered in this chapter.

The Disappearance of the Body

The year 1832 marked the end of the age of spectacular post-mortem punishment of the body. After the gibbetings of William Jobling and James Cook that summer, nobody was hung in chains again in Britain. The same year, the Anatomy Act put an end to the punitive, public dissection of criminals, as the bodies of paupers replaced those of malefactors on the dissection table. The corpses of executed criminals would henceforward remain behind the prison walls, buried in a plain and often unmarked grave in the prison burial ground. This transformation in punishment represents a move towards concealing the body from public view, a trend that eventually relocated the execution itself to the private space of the prison and excluded the community.[6] The sequestration of bodily punishment can be placed alongside other narratives of bodily privacy, including the trend towards specialised private spaces for sleeping and washing, the medicalisation of birth and death, and their abstraction from the places of everyday life, and increasingly anxious discourses about sexuality. Prison execution also demonstrates the end of geographically localised punishment. Hanging at the scene of crime was an important part of the eighteenth-century moral economy. In England, this practice was already in decline in the few decades before 1800, although it continued for longer in Scotland.[7] However, in England outside London, bodies continued to be gibbeted at the scene of crime even when they were executed many miles away, right up until 1832. Whereas hanging in chains at the scene of crime was a strongly community-based punishment, making use of a meaningful location and in turn ensuring that the location remains meaningful in local knowledge, burial inside the prison relocates the body to a "non-place",[8] a space controlled entirely by the State and beyond the reach or experience of the local community.

Hanging in Chains as Deterrent, Retribution or Social Revenge

The wording of the Murder Act suggests that gibbeting was a public act of sanction: "some further mark of infamy". It was thus retributive in nature. As Radzinowicz points out, a recurrent theme in foreigners' accounts of English justice in the eighteenth century was the harsh nature

---

[6] David Cooper (1990) 'Public executions in Victorian England: a reform adrift' in W. Thesing (ed.) *Executions and the British experience from the seventeenth to the twentieth century* (Jefferson, N.C.: McFarland) suggests that public execution would have been abolished far sooner were it not for opposition from radicals who wanted full abolition of the death penalty.

[7] Steve Poole 2015 '"For the benefit of example": processing the condemned to the scene of their crime in England, 1720–1830' in R. Ward (ed.) *A Global History of Execution and the Criminal Corpse* (Basingstoke: Palgrave). For a discussion of Scottish scene-of-crime executions, see Rachel Bennett, unpublished PhD thesis, University of Leicester 2015.

[8] Although I have used Marc Augé's term here, I do not mean 'non-lieux' as he defines them in the sense of being ephemeral places of super-modernity, but in the sense that a non-place "creates neither singular identity nor relations; only solitude, and similitude" (Marc Augé (1995) *Non-Places: introduction to an anthropology of supermodernity* (London: Verso), p. 103).

of the criminal code and the severity of punishments to which the convicted were subject.[9] Post-mortem punishment was thus a result of spiralling inflation of punishment. Extending the kinds of sanction available at the severe end increased the range of possibility and provided a way to distinguish murder or major crimes against the State from less heinous property crimes and crimes against the person which might also result in a sentence of death. As has been noted, when a conviction for damaging the banks of a canal or writing poison pen letters could result in execution[10] (though it rarely did), some visible and striking sanction for the most serious crimes needed to find a way of being worse than death. This could take the form of a particularly painful or horrific execution, punishment of relatives and associates, or further punishment of the corpse. The first two are also known, although by the eighteenth century bloodthirsty executions were reserved for traitors and often were modified or ignored in deference to changing sensibilities; and the punishment of family members, such as the confiscation of property from the heirs of suicides, was also perceived to be against natural justice and widely circumvented.[11] Post-mortem punishment represented a rational response exploiting an irrational but almost universal anxiety among the British people of the period about the proper treatment of the dead body.

We can also conclude that punishment which kept the body from normal churchyard burial was intended to be terrible and horrific. Despite the insistence of Protestant theologians on the insignificance of the dead body, and their strong claims that Christian resurrection did not require a whole and unmutilated corpse, the care taken to present a whole, beautiful body for burial only increased from the seventeenth to the nineteenth century. Several trends relating to the care of the dead body over this period are witnessed in the extensive archaeological evidence; these trends include the change from burial in a winding sheet only to the use of a coffin, beautification of the body using hair pieces, wigs, queues, dentures, special grave clothes and decorating the body with flowers

---

[9] Leon Radzinowicz (1990) *History of English Criminal Law and Its Administration from 1750, volume 1: The movement for reform* (London: Stevens and Sons), pp. 699–720.

[10] D. Hay, 'Property, authority and the criminal law' in Douglas Hay, Peter Linebaugh, John Rule, E P Thompson and Calvin Winslow, *Albion's Fatal Tree: crime and society in eighteenth-century England* (New York: Pantheon books), p. 17.

[11] Macdonald and Murphy, *Sleepless Souls*.

and plants.¹² There is sufficient historical evidence to conclude that the prospect of dissection or hanging in chains was indeed a potent source of dread to the condemned criminal, given the numerous accounts of criminals hearing the pronouncement of their death sentence with equanimity only to fall apart when told that their body would not be returned for burial. When Lambert Reading, for example, was convicted at Chelmsford Assizes in 1775 and sentenced to hang in chains, he begged for that part of his sentence to be revoked in exchange for information about other criminal plans to which he was privy. His request was granted.¹³ Similar accounts of hitherto stoical men collapsing at the horror of being measured or fitted for their gibbet cage are also fairly common. In 1749, Joseph Abseny, a Swedish Catholic condemned for the murder of a servant girl in Bristol, was more troubled by the gibbet part of his sentence than any other and claimed "he did not care if they quarter'd his body so that it was not hung up in the air for Prey to the Birds". Eight years later, John Gatward tried to have the post-mortem part of his sentence altered, although we know from other sources that he was not successful.¹⁴ Only the hardest of criminals had the *sang-froid* to quip, as in 1800 James Wheldon the Lancashire mail-robber did on hearing that his body was to be hung in chains, that he was to be thus "made Overseer of the Highways".¹⁵ Whether this represents a significant change in beliefs about the dead body between the 1750s and the end of the eighteenth century or simply the different attitudes and personalities of the men involved is hard to say. After Theodore Gardelle was hung in chains on Hounslow Heath in 1761, *Read's Weekly Journal* expressed the view that gibbeting "may appear to some people not as an increase but a mitigation of the punishment as, probably, the dread of

---

¹²Sarah Tarlow (1999) *Bereavement and commemoration: an archaeology of mortality* (Oxford: Blackwell); Sarah Tarlow (2011), *Ritual, belief and the dead in early modern Britain and Ireland* (Cambridge: Cambridge University Press); Annia Cherryson, Zoë Crossland and Sarah Tarlow (2012) *A fine and private place: the archaeology of death and burial in post-medieval Britain and Ireland* (Leicester: Leicester Archaeological Monographs).

¹³*London Chronicle*, 5–8 August 1775, issue 2912.

¹⁴*London Evening Post*, 26–29 August 1749; 14–16 April 1757, issue 4593.

¹⁵*The Hull Packet*, 29 April 1800, issue 677.

being antomized (as the vulgar term it) has more effect upon the uninformed mind than that of being exposed upon a gibbet". The journal goes on to recommend that remains be gibbeted only after having first been anatomised, which would be a more terrifying prospect and an enduring example to others.[16]

Whether the threat of the knife or the gibbet was sufficient to prevent serious crime in the first place is also unproven. The murder of Jane Grant by Hannah Bocking, committed during a visit to the Derbyshire gibbet of Anthony Lingard in 1818, suggests that a public example of the consequences of murder was not always an effective deterrent. We have found 21 cases of highway robbery committed close to gibbets in the newspapers, one of coining and one of murder. In fact, on the roads into London, robberies right by the gibbets on Hounslow Heath and Wimbledon Common were repeatedly perpetrated in the later eighteenth century, according to newspaper reports.

However, neither its power as a public and cultural statement of retribution nor its intended deterrent effect distinguishes hanging in chains from the alternative of dissection. The question remains: why gibbet anyone?

Perhaps looking at those instances where crimes that did not come into the purview of the Murder Act were punished by hanging in chains can shed light on the meaning of this punishment. Seventy-eight people in England and Wales were sentenced to hang in chains for crimes other than murder between 1752 and 1834. Sixty-four of those were convicted of mail robbery, highway robbery or Admiralty offences other than murder (such as piracy or mutiny). Table 1.2 summarises the crimes punished by hanging in chains in this period. Notably, apart from murder, crimes that threaten the orderly running of the State, such as interfering with the mail, seem to have particularly merited the especially ostentatious punishment of hanging in chains.

Gibbeting is both more public and more location-specific than dissection. Although the public were allowed to view the opened body of the dissected, their window of opportunity was quite limited, in both time and space. The constricted and controlled space of the dissection room could not admit the vast crowds that typically attended a gibbeting, and the gates and doors of the dissection room allowed the crowd to be filtered

---

[16] *Read's Weekly Journal*, 18 April 1761.

by age, class and gender.[17] By contrast, the gibbet, typically on common land by the public road, enabled the formation of enormous and unregulated crowds. And their longevity meant that anyone who could not visit on the first day could come on the second, or the following week, month, or even years later. When Mary Hardy of Norfolk visited North Yorkshire in 1765, she made a special trip to see the gibbet of Eugene Aram in Knaresborough 16 years after it had been erected there.[18]

The location of the gibbet, unlike the place of dissection, was rarely determined by practicality or custom. Outside London, where gibbets were placed in the same few locations because of the high numbers involved, and some prominent locations near ports or along the Thames estuary where those condemned for maritime crimes were gibbeted according to Admiralty tradition, most gibbets were carefully located at a site that was close to the scene of crime and highly visible, especially from the public road. The lasting and public nature of hanging in chains as a post-mortem punishment, then, seems to have been considered especially appropriate as a response to crimes that outraged the social contract between citizen and State. The security of the national infrastructure—roads, mail, free trade—was defended by the most visible and exemplary of punishments.

THE BODY IN CHAINS

Understanding hanging in chains raises the question: what kind of thing is a gibbeted body? Is it a person? Or is it a thing? Mary Leighton has suggested that dead bodies and human remains occupy an ambiguous position between person and object.[19] The dead body in early modernity was both a person and a thing: this division aligns with a number of other dualisms relating to the newly dead: sentient/not-sentient, powerful/abject, individual/generic.

The gibbeted body is some sort of thing that can be displayed, used in a discourse of power, and in an affective narrative. It may be that

---

[17] Elizabeth Hurren (forthcoming) 'Time, Spectatorship and the Criminal Corpse in eighteenth and nineteenth century England', *Comparative Studies in History and Society*.

[18] *Mary Hardy's Diary* (ed. B. Cozens-Hardy), 1938 (Norfolk Record Society, Vol. 37).

[19] Mary Leighton (2010) 'Personifying objects/ objectifying people: handling questions of mortality and materiality through the archaeological body', *Ethnos* 75(1): 78–101.

the power of the body to represent a whole biography, a whole self, was especially great in the case of the criminal classes: "their bodies are themselves", wrote one eighteenth-century commentator.[20] The body in chains is both an abhorrent and a compelling thing. It is part of a life story and—in its landscape—is the medium through which the story is written. It is narrative, moral and illustration.

Zoë Crossland describes four ways in which the dead body is evidence, or rather, four things that the dead body might be evidence of. These are:

1. Body as evidence of the physical or moral state of the individual
2. Body as evidence of the past (in archaeology, for example, or the religious history of relics)
3. Body as evidence of crime, and
4. Body as evidence of the identity of the person.

To understand that fourth category—the dead body as evidence of identity—Crossland turns to Peircean semiotics to ask what kind of sign the dead body is.[21] Philosopher Kieran Cashell considers that because the dead are absent, signs of the dead are necessarily indexical—holding a relationship with the thing for which they stand.[22] Thus, for example, the identity of a dead parent is signified, indexically, in her engagement ring. But the body *itself* is more iconic than indexical—a relationship based on resemblance rather than association. Thus, the dead body itself can stand for its formerly vital counterpart—the living person. Jeremy Bentham conceived the idea that memorials of the dead could take the form of their own preserved and mounted corpses—a form he called the auto-icon.[23]

---

[20] He was arguing that more brutal physical punishments would make the most effective deterrents.

[21] Zoë Crossland (2009) 'Of clues and Signs: the dead body and its evidential traces', *American Anthropologist* 111 (1): 69–80.

[22] Kieran Cashell (2007) 'Ex post-facto: Peirce and the living Signs of the Dead', *Transactions of the Charles S. Peirce Society* 43(2): 345–71.

[23] Jeremy Bentham attached a paper he had written on the principle of the auto-icon to his own will. He left instructions that his own body was to be prepared in that way (F. Rosen, 'Bentham, Jeremy (1748–1832)', *Oxford Dictionary of National Biography* 2004; online edn., May 2014, http://dx.doi.org/10.1093/ref:odnb/2153).

The gibbeted criminal was a kind of auto-icon, the dead body standing as an iconic sign for the living one. This can work only because the dead body is not the same as the living one—because its deadness moves it at least some of the way from being a person to being a thing.

The executed body is therefore a material thing. Dead bodies in this period are interpreted in different ways according to contexts of discourse. In some discourses, it is de-individualised, universal, medical; in others, it is a highly personalised individual—a signifier of a life. These two tendencies have different historiographies and different meanings and perhaps also constitute the major difference between the alternative post-execution punishments of anatomical dissection and hanging in chains.

Dissection values the body for its universal and biological properties. It anonymises and dislocates the body—gives it no *place*, no material being; de-personalises and ultimately annihilates it. By contrast, hanging in chains proclaims its individual identity and its particular notoriety. Dressed in its own clothes, preserved with "tar" and displayed on a ten-metre post, the gibbeted criminal becomes an enduring spectacle, a warning and a past of local history. Gibbeting a body transforms the criminal into his own memorial and a mnemonic of his crime. Continually encountered by men and women making ordinary journeys, conspicuous standing gibbets ensured that the stories of those criminals would be remembered and retold. Gibbeting creates a memory that will stick in the minds of witnesses and of everyone who hears about it. The mechanisms by which such a memory is created are five:

1. Gibbeting is relatively unusual, infrequent (averaging only one or two a decade in most counties) and out of the ordinary, so that each occurrence is highly individual and distinct.
2. Associated sensory experiences—the smell of the body, the taste of special holiday treats being sold at the gibbet's foot, the sound of a crowd of thousands or, later, of the creaking chains swinging in the wind. All these things contribute to an embodied and fully sensual memory.
3. By associating the gibbet with conspicuous places in the landscape and ensuring its continued visibility. Siting a memorial on a natural eminence beside a well-used road ensures that it is regularly remarked on and discussed, just as one necessarily notices and remarks on the monumental Gormley sculpture "The Angel of the North" every time one passes its prominent location next to the A1 at Gateshead.

4. Ensuring that witnessing the gibbet is a shared experience. It was made possible for huge crowds of people to attend, so that ongoing discussion and reminiscence kept the memory alive.
5. The occasion of gibbeting was made a shared, public event. By ensuring that the original gibbeting is scheduled and time-bound, it becomes an event to be looked-forward to and then to be reminisced about in company.

Dissection and hanging in chains thus represent two very different strategies for dealing with the body of the dead. Although modern commentators frequently allege that post-mortem punishment drew its force from a belief in the necessity of having a whole body for resurrection, theology of the time does not back this up and it is hard to find any contemporary anxiety expressed on this point. As Richardson has noted, such a view is normally quoted only by contemporary supporters of dissection as a way of mocking the ignorance of its opponents.[24] However, there is no doubt that, at a popular level at least, considerable emotional importance was attached to treating the dead body "properly" and "decently"— meaning careful laying out and graveyard burial.

Both punishments denied the possibility of decent burial and thus are attacks on the body. Equally, both are aimed in some way at affecting secular posterity—the memory of the deceased—although this operated very differently between the two. Dissection, although it had its origins in the demands of medicine rather than in jurisprudence, acted as a form of *damnatio memoriae*—a way of obliterating the secular posterity of the individual, of stripping away of personhood by reducing the individual to a type—a human body whose specific or idiosyncratic features were of less interest than its ability to stand for a generic and universal medical body. It is an act of active forgetting. The normal ways of marking a death involve fixing the body by making it beautiful, laying it out in a coffin and committing it to a grave. During the eighteenth century, moreover, there is a growing expectation that a burial plot should belong to the interred in perpetuity, that their body should not normally be moved as was common early modern practice.[25] Dissection takes away

---

[24] Ruth Richardson *Death, dissection and the destitute.*
[25] Sarah Tarlow *Bereavement and Commemoration.*

this possibility and gives the body no place to be remembered and inhibits the creation of a beautiful memory for those left behind.

Hanging in chains also affected memory, but differently. Rather than trying to erase the memory of the condemned, it made that memory notorious and inescapable. Whereas anatomised bodies must be de-personalised, divested of individual biography in order to be useful as a teaching aid (in fact, de-personalisation of the body is essential to the medical practitioner's capacity to maintain "clinical distance"), the gibbeted body needs to hang onto its personal narrative to work its full didactic power. Unlike the dissected body, the gibbeted one must retain its individual personhood. It cannot be universal or generic. Where modern technologies of science helped to create a Foucauldian "medical gaze" of powerful bioscience,[26] the technology of the gibbet facilitated a gaze that shared the theatricality of contemporary anatomy but effected a more personalised and narrative politics of power.

CRIMINAL TALES AND NARRATIVE PERSONS

One key feature of the gibbeted body, then, is that it possesses narrative. If the body is a sign or index of the person, the manner of its death/treatment after death completes a narrative of that person. Narratives are never straightforward. They are created and contested to promote particular interests, and the material signs through which narratives are formed can be carefully deployed in attempts to regulate history.

In the case of the gibbeted body, however, the possible interpretations of the signs deployed in the creation of a dominant narrative of power (i.e. the carefully choreographed display of the body by the State to make a statement about the consequences of refusing to obey the law) easily exceed and subvert attempts to make a single dominant narrative. The political and legal Establishment—insofar as there was any consensus—hoped that the sober contemplation of a gibbeted body would make a primarily moral impression on the crowd, re-enforcing a message about the consequences of serious crime. But as we have seen (Chap. 2, "The Carnival of the Gibbet"), other responses were at least equally present. Because the gibbeted body's treatment is so profoundly different

---

[26] S.R. Kaufman and M. Morgan 2005. "The anthropology of the beginnings and ends of life". *Annual Review of Anthropology* 34: 317–41.

from the treatment of an ordinary dead body, there is some ambivalence in participating as a spectator or crowd member in the carnival of the gibbet. It is thrilling to see, to be physically close to, a dangerous criminal body, but it is also transgressive.

The glamorous appeal of the criminal dead is built on the fame or notoriety of the individual whose body it is/was and the attendant thrill of danger. An encounter with the living body of a criminal—especially a violent one—is dangerous. The potential risk of physical harm, however, is tamed by execution. The criminal corpse still looks like the thing it was but is rendered inert—harmless—by death. The family picnic under the gibbet is analogous to the photograph of a grinning hunter with his foot on the neck of a dead lion: an easy claim to bravery and a bid for contagious glamour.

The gibbeted body is both less and more than a dead human being. It is less because the richness of experience, the animation of life is gone; more because it has acquired symbolic properties that were never present before.

Conclusions: Hanging in Chains

From the review of hanging in chains covered in this chapter, a few notable elements emerge: first, it was a comparatively infrequent element of punishment. Even when a sentence contained a post-mortem element, whether mandatory or discretionary, it was far more likely to be dissection than gibbeting. Less than 15% of all crimes falling under the terms of the Murder Act were punished by hanging in chains. Second, it was very expensive. The costs of gibbeting a single criminal could exceed a year's pay for a labouring man.[27] Third, despite or perhaps because of their infrequency, gibbetings were of huge public interest and often were attended by tens of thousands of people who would journey considerable distances to witness the body gibbeted. It would be fair to describe the events around the hanging in chains as in some ways carnivalesque, when vast crowds were provided with food, drink and entertainment. The magnitude of the event was one thing that made the event and the criminal highly

---

[27] Assuming a day rate of around a shilling a day for 6 days a week and a working year of around 50 weeks.

memorable and significant in local minds. Gibbets were also remembered in the longer term through toponyms, stories and the curation of the material gibbet itself, which could remain in situ for many decades.

What are we to make of hanging in chains? It is an unusual punishment and feels, in the later eighteenth and nineteenth centuries, anachronistic. It has the feel of early modern spectacular and theatrical punishments visited publicly on the body of the criminal. The gibbet is in some ways the spectacular bodily punishment *par excellence*. The criminal is already dead and the gibbet has no part to play in the actual execution. Nor does it in any way sequester crime from society or protect society from crime. It is pure theatre.

Friedland's recent discussion of Foucault (2012) notes the anthropological intention of Foucault's work on the history of punishment: work that stresses the function of spectacular punishment to be more than an act of terror or an exemplary deterrent.[28] Instead, Foucault drew attention to the capacity of spectacular punishment to be an act of social restitution, a theatrical and ceremonial event that will in some ways knit up the hole in the social fabric that was rent by the crime itself. In his own analysis of spectacular punishment in France, Friedland argues that public executions should be seen more "as meaningful rituals, which allowed the community at large to find redemption ... than as any kind of display of sovereign majesty" (2012: 13).

Many cases of hanging in chains fit well into just such an explanation: gibbetings, unlike dissections, usually were carried out close to the scene of crime, which tended to be in the murderer's and often the victim's own community, and had the pleasing symmetry that the crime and its punishment happened in the same place.

However, the more anthropological view of spectacular punishment as communal restitution does not mean that it was not also a declaration of State power and a means of negotiating the relationships of power and control by which eighteenth-century British society was structured. Because it was both spectacular and infrequent, the currency of hanging in chains

---

[28] Paul Friedland (2012) 'Introduction: reading and writing a history of punishment' In *Seeing Justice Done: the age of spectacular capital punishment in France* (Oxford: Oxford University Press), pp. 11–14. Friedland also notes that Foucault's actual discussion does not fully support this position, instead emphasising terror and the need of the State to make a show of their repressive power.

was high. As a demonstration of State power, it certainly would have been highly visible. What such a demonstration accomplished is harder to say.

Ultimately, it is too simplistic to oppose interpretations of the gibbet as *either* an uncontested demonstration of State power *or* a communal ritual of popular justice or subaltern subversion. The crowd attending a gibbeting was a diverse body, and whereas some were undoubtedly appalled by the brutality, others were undoubtedly impressed by the moral lesson or titillated by the close encounter with criminal glamour and violent death. Attempts to subvert a State-scripted theatre of power certainly took place but make sense only in a context in which the demonstration of force could normally be expected to make a strong emotional impact.

THE POWER OF HANGING IN CHAINS

Attitudes towards the gibbet are complex and contradictory. There is a tension between disgust and revulsion on the one hand and fascination on the other. This tension is still evident in the context of contemporary interest in crime history. Brutal physical punishment of the body excites far more public interest than, for example, the history of tax law or boundary disputes. Where gibbet cages survive in museum collections, nearly all are on display and many are among the most popular visitor attractions. The gibbet cages at Moyses Hall, Bury St Edmunds and South Shields Museum are located in the main downstairs galleries, close to the entrance, and visitors to Nottingham's Galleries of Justice encounter a gibbet cage hanging from the ceiling of the atrium.

The popularity of these exhibitions and displays relates in large measure to the taboo-busting power of making visible the invisible interior of the body or of illuminating the normally secret and hidden process of bodily decay. In connection with the first of these, we might note the record-breaking commercial success of Gunther von Hagens's "Body Worlds" phenomenon; in connection with the second, the number of peak-time television dramas that feature prominently the work of forensic scientists examining the taphonomic processes at work on a cadaver.[29]

---

[29] Since the beginning of the Body Worlds exhibition series in 1995, featuring preserved "plastinated" bodies posed to demonstrate their organs and structure, over 40 million visitors in more than 90 cities worldwide have visited a Body Worlds exhibition (http://www.bodyworlds.com/en/exhibitions/questions_answers.html, accessed 15/6/2015).

But beyond the interest in dead bodies generally, is there particular power attached to the dead criminal body? The criminal corpse is culturally located in the overlap between crime and dead bodies. Both of these areas are deviant, hidden, non-normative. The double dose of transgressive and normally sequestered areas of experience is very effective in stimulating the prurient interests of the public. There were practical reasons for locating gibbets on marginal land, but such places were also symbolically appropriate for the liminal criminal corpse: dead but not buried; a person transformed into a thing; existing, but not living.

Post-mortem punishment gained power through its distance from normal burial and funerary rites. As an ostentatious exclusion from normative mortuary practices, post-mortem punishment ensured that the desired "respectable" and "decent" end was out of the question for criminals. In the period leading up to the Murder Act, it had become a common custom for those criminals able to afford the expense to arrange their own transport to the place of execution in a mourning coach of the kind more usually associated with respectable funerals. Comment in the press opposed this practice on the grounds that it was contrary to the ends of "Ignominy and Shame" which should properly attend such an occasion.[30]

Hanging in chains was an attempt—after the peak period of spectacular bodily punishment—to shame and humiliate the bodies of the most serious criminals. By "making an example" in a carefully choreographed way, the Establishment intended to enforce social conformity in respect of law. But given the polyvalence of the dead body, attempts to produce a certain narrative of crime were never fully regulated. The gibbeted body could be recruited into a number of other stories with a different moral value, including the implication that the State itself was demeaned and barbarous to use such a disgusting and unsubtle punishment. Eventually, the practice had all but died out many years before its final abolition in 1834.

Hanging in chains, then, was too brutal, ultimately, for the more educated and progressive elements of nineteenth-century British society to be comfortable with. What is more, it had arguably never proved an adequate tool of social control, because the multiple narratives of the

---

[30] *London Evening Post*, 25 September 1750. As Wilf notes, "The aesthetics of mourning, centered around themes of dignity and honor, undercut the stigma of a public hanging" ('Imagining Justice', p. 58).

criminal corpse were never contained. Instead, the powerful criminal corpse maintained the capacity to subvert or twist any official attempt to harness its power.

Open Access This chapter is licensed under the terms of the Creative Commons Attribution 4.0 International License (http://creativecommons.org/licenses/by/4.0/), which permits use, sharing, adaptation, distribution and reproduction in any medium or format, as long as you give appropriate credit to the original author(s) and the source, provide a link to the Creative Commons license and indicate if changes were made.

The images or other third party material in this chapter are included in the chapter's Creative Commons license, unless indicated otherwise in a credit line to the material. If material is not included in the chapter's Creative Commons license and your intended use is not permitted by statutory regulation or exceeds the permitted use, you will need to obtain permission directly from the copyright holder.

Appendix 1: All Cases of Hanging in Chains

Name of offender	County	Gibbet location	Date of sentencing	Date of execution
Edward Tool	Middlesex	Finchley Common	15/02/1700	02/02/1700
Michael Van Berghen and Dromelius Beachere	Middlesex	Between Mile End and Bow	24/06/1700	19/07/1700
William Elby	Middlesex	Fulham	02/08/1717	12/09/1707
Herman Brian	Middlesex	Acton Gravel Pits	16/10/1707	24/10/1707
William Johnson	Middlesex	Near Holloway, between Islington and Higate	10/09/1712	19/09/1712
Richard Keele and William Lowther	Middlesex	Holloway	10/12/1713	23/12/1713
John Tomkins	Middlesex	Unknown	27/02/1717	20/03/1717
Joseph Still	Middlesex	Kingsland Road	28/02/1717	22/03/1717
John Price	Middlesex	Stone Bridge by Kingsland	24/04/1718	31/05/1718
Matthew Clark	Middlesex	Wilsden Green	Unknown	28/07/1721

(continued)

APPENDIX 1: ALL CASES OF HANGING IN CHAINS

Name of offender	County	Gibbet location	Date of sentencing	Date of execution
George Simpson, John Hawkins and Benjamin Child	Middlesex	Hounslow Heath	10/05/1722	21/05/1722
James Shaw	Middlesex	Kentish Town	12/01/1722	08/02/1722
Robert and William Bolas	Shropshire	Bolas Hole, River Tern	Unknown	Unknown
Rice Jones	Denbighshire	Wrexham	10/04/1726	15/04/1726
John Gutteridge	Surrey	Bristow Causeway	19/03/1724	01/04/1724
John Humphrey	Glam	Unknown	05/09/1726	07/10/1726
George Cutler and John Winter	Hampshire	Waltham Chase	11/03/1726	11/03/1726
Thomas Billings	Middlesex	100 yards from where hanged, executed at Tyburn	20/04/1726	09/05/1726
Edward Burnworth and William Blewitt	Middlesex	St Georges Fields, over 'the two fighting cocks' in the mint- Mint is in Southwark	05/04/1726	10/04/1726?
Emanuel Dickinson and Thomas Berry	Middlesex	Kennington Common (gallows where St Mark's Church is)	05/04/1726	10/04/1726?
Legee and John Higgs	Middlesex	Putney Common (Putney Heath)	05/04/1726	10/04/1726?
Roger Bryany	Gloucs	Unknown	11/03/1727	Unknown
Henry Brookman	Somerset	Belton Meeting House, Hursley Hill	08/04/1727	26/04/1727
William 'Old' Skull	Somerset	Old Down	14/08/1729	Unknown
John Wilson	Surrey	Kennington Common (gallows where St Mark's Church is)	08/04/1729	Unknown
Ferdinando Shrimpton and Robert Drummond	Middlesex	Stamford Hill (near Joseph Still)	28/02/1730	17/04/1730

(continued)

APPENDIX 1: ALL CASES OF HANGING IN CHAINS   121

Name of offender	County	Gibbet location	Date of sentencing	Date of execution
Hugh Horton (Norton)	Middlesex	Hounslow Heath	08/04/1730	12/05/1730
Robert Weaver	Herefordshire	White Hill, Weobley	14/03/1731	27/03/1731
John Chappel	Middlesex	Stone Bridge	24/02/1731	08/03/1731
William Williams	Middlesex	Turnham Green	24/02/1731	08/03/1731
John Naden	Staffs	Gun-Heath, Leek	16/08/1731	30/08/1731
Benjamin Cruse and Stephen Woon	Devon	Unknown	20/03/1732	12/04/1732
Ely Hatton	Gloucs	Meane Hill near Mitchel Dean	19/08/1732	4/09/1732
Isaac Hallam	Lincolnshire	Hanged at Nettleham, gibbeted near crime, buried in the church?	05/03/1733	20/03/1733
Thomas Hallam	Lincolnshire	Hanged at Faldingworth, where also gibbeted	05/03/1733	20/03/1733
John Notton	Suffolk	Rymerton	05/03/1734	03/04/1734
John Jacob Davies	Sussex	Ditchelling Common	12/08/1734	21/08/1734
Herbert Hayns	Essex	Essex	23/07/1735	08/08/1735
Edmund Goodrich	Gloucs	Corse Lawn, Cheltenham	9/08/1735	22/08/1735
Samuel Gregory	Middlesex	Edgeware Road (runs by St Georges Fields)	26/02/1735	04/06/1735
William Blackwell	Middlesex	Near Paddington	15/10/1735	08/11/1735
Joseph Rose, William Bush, Humphry Walker, John Field	Middlesex	Near Edgeware	26/02/1735	10/03/1735
Evan Hugh Jones	Montgom	Manafon	02/08/1735	??/August
John Weekes	Sussex	Unknown	02/08/1735	11/08/1735

(continued)

122 APPENDIX 1: ALL CASES OF HANGING IN CHAINS

Name of offender	County	Gibbet location	Date of sentencing	Date of execution
Thomas and Richard Marshall	Bucks	Rye Common, Chipping Wicombe	8/03/1736	22/03/1736
Hugh Moss and William Hawthorne	Cheshire	Ettley Heath	21/04/1736	08/05/1736
David Anderson	Kent	Hambledown	16/08/1736	2/09/1736
John and Joseph Emerson	Surrey	Kennington Common	6/08/1735	20/08/1735
Thomas savage	Warwickshire	12 miles from Warwick	Unknown	04/1737
Michael Moorey	Hampshire	Arreton, Isle of Wight	02/03/1737	19/03/1737
William Maw and Jeffrey Morat	Middlesex	Shepherd's Bush near Kensington gravel pits (just east of Kensington Gardens)	16/02/1737	03/03/1737
John Sturabout	Berkshire	Coldbourne Hill, Tilehouse Heath	27/02/1738	11/03/1738
George Price	Middlesex	Hounslow Heath?	13/01/1738	21/02/1738
Gill Smith	Surrey	Kennington Common	16/03/1738	10/04/1738
John Cotton	Northants	Paulerspury Common	06/03/1739	22/03/1739
James Caldclough and Joseph Morris	Middlesex	Hounslow Heath	07/06/1739	02/07/1739
Michael Curry	Northumbs	St Mary's Isle	20/08/1739	4/09/1739
Thomas Limpous	Somerset	Dunkit Hill, mile from Wells	29/08/1739	21/09/1739
Thomas Willot	Staffs	Mere Heath	15/03/1749	04/1739
Benjamin Randall	Bucks	Loudwater	16/07/1740	08/08/1740
Thomas Edwards	Denbighshire	Llangollen	22/08/1740	08/09/1740
Edward Ellis	Flintshire	Flint Marshes	12/04/1740	26/04/1740

(continued)

APPENDIX 1: ALL CASES OF HANGING IN CHAINS   123

Name of offender	County	Gibbet location	Date of sentencing	Date of execution
Cornelius York	Somerset	Brislington Common	31/03/1740	03/05/1740
John Millard	Somerset	Bedminster Down	31/08/1740	04/09/1740
Charles Drew	Suffolk	Long Melford	24/03/1740	9/04/1740
William Creake	Surrey	Gibbet Lane between Camberley and Bagshot	30/07/1740	25/08/1740
Henry Wheeler	Wiltshire	Unknown	08/03/1740	29/03/1740
Bryan Connell	Northants	Weedon Common	10/03/1741	03/04/1741
Matthew Mahony and Captain Goodere	Somerset	Bristol, by the river	02/04/1741	17/04/1741
John Carr	Middlesex	Finchley Common	05/04/1741	02/05/1741
James Hall	Middlesex	Shepherd's Bush, 3 miles from Tyburn Turnpike, on Road to Acton	28/08/1741	14/09/1741
Lawrence Holliday	Sussex	Fairlight Common	16/03/1741	01/04/1741
Richard Pilgrim	Herts	Knebworth	04/03/1742	22/03/1742
William Tyler	Lincolnshire	Pinchbeck Drainpipe	Unknown	03/1742
Robert Carleton	Norfolk	Diss	18/03/1742	05/04/1742
Joseph Mulloe	Gloucs	Rodborough Hill	03/03/1743	22/03/1743
Domingo Goodheart	Hampshire	Unknown	01/03/1743	16/03/1743
John Roberts and Hugh Edward	Caernarfon	Twllhely Marsh	Unknown	22/04/1743
John Breads	Sussex-Rye	Rye	Unknown	08/06/1743
John Codlin	Norfolk	Bunwell	10/03/1743	07/04/1743
Edward Wollaston	Shrops	Unknown	30/07/1743	08/1743
John Knott	Beds	Luton Down	12/07/1744	28/07/1744
Andrew Burnet and Henry Payne	Gloucs	Hot Well, Durdham Down	03/03/1744	22/03/1744
Thomas Cambrey	Gloucs	Bowling Green House, Cirencester	03/03/1744	20/03/1744

(continued)

124 APPENDIX 1: ALL CASES OF HANGING IN CHAINS

Name of offender	County	Gibbet location	Date of sentencing	Date of execution
John Snell	Herts	Hanging Wood	28/03/1745	03/1745
Thomas Dyer	Devon	Unknown	17/03/1746	11/04/1746
John Parr	Oxfordshire	Banbury	Unknown	Unknown
Matthew Henderson	Middlesex	Edworth Road (possibly Edgware Road?)	9-11/04/1746	25/04/1746
Francis Wilkins	Somerset	Black Down	2/09/1746	09/1746
Marey John Galway	Somerset	Chiluton Heath (Chilton Cantilo?)	2/09/1746	09/1746
Samuel Hullock	Middlesex	Stamford Hill	Unknown	31/07/1747
Richard Ashcroft	Middlesex	Shepherd's Bush	04/06/1747	16/07/1747
John Cook	Middlesex	Shepherd's Bush	15/07/1747	16/07/1747
Hosea Youell	Middlesex	Stamford Hill	14/10/1747	16/11/1747
Samuel Austin	Middlesex	Shepherd's Bush	09/12/1747	21/12/1747
Adam Graham	Cumberland	Kingmoor, Carlisle	13/08/1748	10/09/1748
Thomas Bibby	Herts	Gravel Pits, St Albans	10/03/1748	25/03/1748
Francis Herbert	Kent	Unknown	21/03/1748	14/04/1748
William Hartnup	Kent	Goudhurst Gore	21/03/1748	14/04/1748
John Juckers	Cambs-Ely	Whitlesea	Unknown	07/11/1748
Arthur Gray	Middlesex	Stamford Hill	20/04/1748	11/05/1748
William Whurrier	Middlesex	Finchley Common	24/02/1748	18/03/1748
William Salter	Norfolk	Holt	10/03/1748	12/04/1748
Richard Biggs	Somerset	Three Holes Down, Bath	23/08/1748	14/09/1748
Stephen Pettitt	Suffolk	A mile from Ipswich	14/03/1748	02/04/1748
Abraham Durrill	Wilts	Great Bedwyn	05/03/1748	30/03/1748
Joseph Abseny	Gloucs	Durdham Down	05/08/1749	25/08/1749
Thomas Kingsmill	Middlesex	Goudhurst Gore	05/04/1749	26/04/1749
William Fairall	Middlesex	Horsmonden (Gibbet Lane)	05/04/1749	26/04/1749
Richard Maplesden	Middlesex	Lamberhurst or Lewes, Sussex	05/07/1749	04/08/1749
James Watkins	Monmouths	Unknown	09/03/1749	29/03/1749

(continued)

APPENDIX 1: ALL CASES OF HANGING IN CHAINS   125

Name of offender	County	Gibbet location	Date of sentencing	Date of execution
Thomas Supple	Surrey	Kingston	03/08/1749	25/08/1749
Benjamin Tapner	Sussex	Rook's Hill near Chichester	16/01/1749	18/01/1749
William Carter	Sussex	Near Rake	16/01/1749	18/01/1749
John Cobby and John Hammond	Sussex	Selsey Isle	16/01/1749	18/01/1749
John Mills	Sussex	Sindon Common	13/03/1749	20/03/1749
Henry Shearman	Sussex	Rake	13/03/1749	21/03/1749
Edmund Richards	Sussex	Hambrook Common	29/07/1749	9/08/1749
George Chapman	Sussex	Hurst Common	29/07/1749	19/08/1749
Gabriel Tomkins	Beds	Chalk Hill between Dunstable and Hockley	8/03/1750	23/03/1750
Garrett Delaney and Edward Johnson	Cheshire	Great Saughall	03/09/1750	22/09/1750
Thomas Nunn, John Hall	Essex	Harrolds Wood Common, near Rumford Gallows	13/03/1750	6/04/1750
Richard Merrick	Gloucs	By the Monument on Lansdown	10/03/1750	28/03/1750
William Kemp	Hampshire	Unknown	06/03/1750	16/06/1750
John Ogleby	Kent	Alberry Hill	31/07/1750	23/08/1750
John Barchard	Norfolk	By the sea, 1 mile from Yarmouth Gallows	10/08/1750	26/09/1750
Thomas Wakelin	Northants	Dunstable Rd	13/03/1750	23/03/1750
Toby Gill	Suffolk	Blythburgh	16/08/1750	14/9/1750
James Cooper	Surrey	Croomhurst	09/08/1750	30/08/1750
Thomas Colley	Herts	Gubblecote	29/07/1751	24/08/1751
James Welch and Thomas Jones	Surrey	Drixton Causeway	15/08/1751	Unknown
Robert Steel	Middlesex	Shepherd's Bush	11/09/1751	23/10/1751

(continued)

126   APPENDIX 1: ALL CASES OF HANGING IN CHAINS

Name of offender	County	Gibbet location	Date of sentencing	Date of execution
John Young (Davy?)	Devon	Ingoldsby Common	16/03/1752	3/04/1752
Anthony Colpris	Dorset	Windmill Point, Poole	23/07/1752	25/07/1752
John Swan	Essex	Ten Mile stone, Epping Forest (Buckets Hill?)	9/03/1752	28/03/1752
John Grace	Kent	Howe/Hope Common near Rochester	29/07/1752	13/08/1752
William Chaplain	Norfolk	Kings Lynn Common	28/07/1752	22/07/1752
John Salisbury	Middlesex	Hounslow Heath/ Smallberry Green	08/04/1752	27/04/1752
Anthony De Rosa	Middlesex	Stamford Hill	19/02/1752	03/1752
Thomas Otley	Suffolk	Black Close Hill near the road to Newmarket	23/07/1752	27/07/1752
Robert Derby	Surrey	Black Water	30/03/1752	24/04/1752
Christopher Johnson and John Stockdale	Middlesex	Winchmore Hill	18/07/1753	23/07/1753
William Morgan	Radnorshire	Llowes	04/04/1754	10/04/1754
George Davies	Bucks	Gerrards Cross/ Holtspur Heath, road leading from Beaconsfield to High Wycombe	10/03/1755	31/03/1755
Josiah Hugh	Glams	Penmark	19/08/1755	10/09/1755
Matthew Snatt	Essex	Buckett's Hill/ Bare Faced Stagg in Epping Forrest	29/07/1757	12/08/1757
Edward Morgan	Glams	Eglwysilan common	30/03/1757	06/04/1757
John Gatward	Herts	Colliers End	11/04/1757	27/04/1757
John Freeman	Cambs-Ely	Guyhirn	Unknown	17/10/1757
James Pookey	Essex	Chinkford Hatch, on Epping Forrest	13/03/1758	18/03/1758
William Moore	Kent	Chatham Hill	20/03/1758	27/03/1758

(continued)

APPENDIX 1: ALL CASES OF HANGING IN CHAINS   127

Name of offender	County	Gibbet location	Date of sentencing	Date of execution
Benjamin Downing	Essex	Radwinter	12/03/1759	30/03/1759
John Grindrod	Lancashire	Pendleton Moor	17/03/1759	24/03/1759
Thomas Brown	Lincolnshire	Ancholm Corner near Spital	12/03/1759	28/03/1759
Thomas and Joseph Darby	Shropshire	Darby's Hill, Oldbury	6/08/1759	11/08/1759
Robert Saxby	Surrey	Wootton	09/08/1759	13/08/1759
Eugene Aram	Yorkshire	Knaresborough Forrest	28/07/1759	6/08/1759
Francis Roper	Glams	Llantwit	09/04/1760	15/04/1760
John Cardinal and Jacob Murton	Herts	Unknown	05/12/1760	17/03/1760
William Odell	Middlesex	Ealing Common (near Acton)	10/09/1760	15/09/1760
Patric McCarty	Middlesex	Finchley Common	22/10/1760	25/10/1760
Robert Williams	Glams	Swansea	04/08/1761	10/08/1761
Francis Arsine	Hampshire	Blockhouse Point	29/06/1761	4/07/1761
Daniel Ginger	Herts	Colney, 15 miles from Hertford	04/03/1761	11/03/1761
Jean Baptiste Pickard	Kent	Sissinghurst?	16/03/1761	25/03/1761
Patrick Ward	Somerset	Broad Pitt near Kingsroad	Unknown	20/10/1761
Theodore Gardelle	Middlesex	Hounslow Heath	01/04/1761	04/04/1761
Richard Parrott	Middlesex	Hounslow Heath	21/10/1761	26/10/1761
Edward Johnson	Suffolk	Sudbury	16/03/1761	23/03/1761
Daniel Ryan	Lincolnshire	Unknown	26/07/1762	08/1762
John Plackett	Middlesex	Finchley Common	14/07/1762	07/1762
William Buckley	Worcs	Wyre Forest	11/08/1762	14/08/1762
George Harger	Yorkshire	Unknown	06/03/1762	18/04/1762
Thomas Hanks	Gloucs	Near where murder committed	9/03/1763	14/03/1763
Daniel Blake	Middlesex	Hounslow Heath	23/02/1763	02/1762

(continued)

APPENDIX 1: ALL CASES OF HANGING IN CHAINS

Name of offender	County	Gibbet location	Date of sentencing	Date of execution
Thomas Watkins	Berks	Gallows Lane, Windsor	5/03/1764	09/03/1764
John Croxford	Northants	Hollowell Green	31/07/1764	4/08/1764
William Corbet	Surrey	Gallery Wall between Rotherhithe and Deptford	29/03/1764	6/04/1764
William Jacques	Wiltshire	Stanton Fields (Stanton St Quentin)	4/08/1764	14/08/1764
Andrew Benevenuto and Simon Pignano	Kent	Pennenden Heath	Unknown	Unknown
Edward Drury, Robert Lesley, Moses Baker	Warwickshire	Stoneleigh Common	6/04/1765	17/04/1765
John Smith	Lancashire	Liverpool, Beacon's Gutter, mile from town	09/08/1766	23/08/1766
William Whittle	Lancashire	Cliff Lane Ends in Farington (40 yards of father-in-law's house, three miles from Preston on Liverpool Road by way of Croston)	29/03/1766	05/04/1766
Phillip Phillip	Camarthen	Newcastle Emlyn	15/04/1767	22/04/1767
Thomas Nicholson	Cumberland	Carlton Fell - Penrith	25/08/1767	31/08/1767
Robert Rymes	Dorset	Western Road	19/03/1767	24/03/1767
Robert Jones	Gloucs	Bourton on the Hill	01/08/1767	07/08/1767
Robert Downs	Notts	Mansfield Forest	06/08/1767	10/08/1767
John Scott	Shropshire	Coppy foot on the Morse, Bridgnorth	4/04/1767	21/04/1767
James Williams	Hampshire	Southsea Beach, Portsmouth	16/07/1768	28/07/1768
John Curtis	Wiltshire	Lower Burn Beck, Britford	5/03/1768	14/03/1768
Thomas Lee	Yorkshire	Grassington Gate (gibbet hill, grass wood)	16/07/1768	25/07/1768

(continued)

APPENDIX 1: ALL CASES OF HANGING IN CHAINS

Name of offender	County	Gibbet location	Date of sentencing	Date of execution
John Whitfield	Cumberland	Near Armithwaite	29/07/1769	9/08/1769
Philip Hooton	Lincolnshire	Surfleet Common	01/03/1769	06/03/1769
John Bowland	Rutland	Empington Common, Empington Corner GNR	07/07/1789	15/07/1789
Robert Hazlett	Durham	Gateshead Fell	14/08/1770	18/09/1770
William Spiggott and William Walter	Herefords	Hardwick Common, near Hay	24/07/1770	30/07/1770
Evan John Stretton	Middlesex	Finchley Common	11/07/1770	01/08/1770
Peter Conoway and Michael Richardson	Middlesex	Bow Common	11/07/1770	01/08/1770
John Franklin	Wiltshire	Bockington Abney	31/03/1770	20/04/1770
John/Jack Uppington	Sussex	Wepham Wood, now called gibbet piece	18/03/1771	6/04/1771
William Keeley	Gloucs	Campden	22/08/1772	28/08/1772
Jos Guyant and Jos Allpress	Middlesex	Finchley Common	03/06/1772	08/07/1772
Edward Corbett	Bucks	Bierton	19/07/1773	23/07/1773
William Amor	Wiltshire	Pewsey Down	06/03/1773	16/03/1773
Walter Kidson	Worc	Stourbridge Common	21/08/1773	27/08/1773
Robert Jones	Flintshire	Lightwood	05/04/1774	25/04/1774
Thomas Owen	Radnorshire	Evenjobb Hill, New Radnor	23/03/1754	29/03/1754
Robert Thomas	Yorkshire	Beacon Hill, Halifax	16/07/1774	6/08/1774
Matthew Cocklane	Derby	Bradshaw Hay	16/03/1776	21/03/1776
Matthew Norminton	Yorkshire	Beacon Hill, Halifax	9/03/1776	15/04/1776
Samuel Thorly	Cheshire	Congleton Heath	3/04/1777	10/04/1777
John Thomas	Denbighshire	Rosset Green, Marfod Mill	28/03/1777	21/04/1777

(continued)

130 APPENDIX 1: ALL CASES OF HANGING IN CHAINS

Name of offender	County	Gibbet location	Date of sentencing	Date of execution
James Hill	Hampshire	Blackhouse Point	04/03/1777	10/03/1777
Joseph Armstrong	Gloucs	Near Cheltenham	Unknown	Unknown
Morris Rowlands	Caernarfon	Dalar Hir	18/04/1778	25/04/1778
Thomas Arthur	Glams	Monydd Buchan	22/08/1778	28/08/1778
Joseph Maseley	Hampshire	Exton	03/03/1778	09/03/1778
John Spencer	Notts	Scrooby	22/07/1779	26/07/1779
George Easthop	Staffordshire	Cradley Heath	25/03/1779	29/03/1779
William Wotton	Devon	Broadbury Down	13/03/1780	20/03/1780
Thomas/John Knight	Kent	Bostall Hill, Whitstable	13/03/1780	18/03/1780
John Andrews	Devon	Unknown	6/08/1781	24/08/1781
John Bryan	Hampshire	Blockhouse Point, near John the Painter's gibbet	24/07/1781	30/07/1781
Thomas Hammond and John Pitmore	Warwickshire	Washwood Heath	27/03/1781	2/04/1781
Charles Storey	Kent	Cartham	22/07/1782	26/07/1782
William Smith	Middlesex	Finchley Common	20/02/1782	24/04/1782
Francis Fearn	Yorkshire	Loxley Common	13/07/1782	23/07/1782
Jenkin William Prothero	Gloucs	Durdham Down	26/03/1783	31/03/1783
William Peare	Wiltshire	Near Chippenham	Unknown	19/08/1783
George Goodwin	Somerset	Bedminster Down	23/08/1783	10/09/1783
James May and Jeremiah Theobald	Suffolk	Eriswell	18/03/1783	24/03/1783
John Clay	Warwicks	Chilvers Common	22/03/1783	29/03/1781
Thomas Wardle	Worc	Bromsgrove Lock	02/08/1783	7/08/1783
Thomas and Henry Dunsden	Gloucs	Shipley-cum-Wichwood	24/07/1784	30/07/1784
Richard Rendall	Somerset	Totterdown	25/03/1784	08/04/1784
James Cliffen	Norfolk	Dereham	17/03/1785	24/03/1785

(continued)

APPENDIX 1: ALL CASES OF HANGING IN CHAINS   131

Name of offender	County	Gibbet location	Date of sentencing	Date of execution
John Roberts	Northants	Northampton	24/2/1785	5/03/1785
John Price	Oxford	Milton Common	02/03/1785	07/03/1785
John Hastings	Hampshire	Hardway, near Gosport	25/07/1786	31/07/1786
Gervaise Matcham	Hunts	Wolley Rd Junction	29/07/1787	2/08/1787
John Shilling	Norfolk	Burnham Thorpe	17/03/1786	25/03/1786
Abraham Tull and William Hawkins	Berks	Mortimer Common	7/03/1787	9/03/1787
John Kennedy and Thomas Smith	Herts	Charley Wood Common	19/07/1787	03/08/1787
Edward Lannigan, James Marshall and Michael Casey	Surrey	Hindhead Common	2/04/1787	7/04/1787
William Emmanuel	Camarthen	Pembrey Common	02/08/1788	09/08/1788
John Richards and William Smith	Devon	Stoke	17/03/1788	24/03/1788
Cornelius Carty	Middlesex	Four Mile Stone, Edgeware Road	14/01/1789	01/1789
Richard and William Weldon	Rutland	Hambleton Hill	13/03/1789	16/03/1789
John Walford	Somerset	Doddington Green, Walford's Gibbet	15/08/1789	20/08/1789
John Dean	Cheshire	Stockport Moor	31/08/1790	2/09/1790
William Saville	Essex	Manuden	8/03/1790	15/03/1790
William Jones	Herefordshire	Longtown Green	29/07/1790	2/08/1790
James Macnamara	Lancashire	Kersal Moor	14/08/1790	11/09/1790
Thomas Jackson	Norfolk	Methwold Common	12/03/1790	19/03/1790
Henry Lowndes	Cheshire	Helsby Tor	14/04/1791	21/04/1791
William Winter	Northumbs	Steng Cross	04/08/1792	10/08/1792
William Anthony	Norfolk	Kettlestone Common	16/03/1792	24/03/1792

(continued)

132   APPENDIX 1: ALL CASES OF HANGING IN CHAINS

Name of offender	County	Gibbet location	Date of sentencing	Date of execution
Ralph Smith	Lincolnshire	Frampton, near where the murder was committed	10/03/1792	16/03/1792
John Day	Middlesex	Kennington Common	Unknown	09/1789
Roger Benstead	Suffolk	Undly Common near Lakenheath	21/03/1792	26/03/1792
Spencer Broughton	Yorks	Broughton Lane Attercliffe	19/03/1792	14/04/1792
Francis Martin	Devon	Brow of Halldown	18/03/1793	28/03/1793
Edward Miles	Lancashire	The Twysters, Manchester Road, Warrington	10/08/1793	14/09/1793
John Bettley, John Riddle and Richard Ellis	Staffordshire	Unknown	14/03/1793	20/03/1793
Edward Howell and James Rook	Sussex	Peterdene Lane, near Shoreham	18/03/1793	23/04/1793
Patrick Quin/ Coine	Hampshire	Northwood, Isle of Wight	29/07/1794	16/08/1794
Francis Jennison and William Butterworth	Hampshire	Cumberland Fort	29/07/1794	04/08/1794
John Nichols	Suffolk	Honington	19/03/1794	26/03/1794
Thomas Campion	Devons	Bovey Tracey	27/07/1795	6/08/1795
Stephen Watson	Norfolk	West Bradenham	20/03/1795	25/03/1795
William Bennington	Norfolk	West Dereham	20/03/1795	25/03/1795
James Scully and Michael Quin	Cambs-Ely	Wisbech	Unknown	24/10/1795
Jeramiah Abershaw	Surrey	Putney Heath	27/07/1795	03/08/1795
Thomas Brown and James Price	Cheshire	Mickle Trafford	4/04/1796	30/04/1796
William Suffolk	Norfolk	North Walsham	17/03/1797	24/03/1797

(continued)

APPENDIX 1: ALL CASES OF HANGING IN CHAINS   133

Name of offender	County	Gibbet location	Date of sentencing	Date of execution
George Prince	Hampshire	Brook, Bramshaw	18/07/1794	23/07/1798
Thomas Austin	Kent	Charing Heath	11/03/1799	28/03/1799
John Haines and Thomas Clarke	Middlesex	Hounslow Heath	09/01/1799	03/1799
Richard Williams	Somerset	Ilton	28/03/1799	01/04/1799
Robert and William Drewitt	Sussex	North Heath Common, near Midhurst	25/03/1799	13/04/1799
John Holt	Berkshire	Curbridge Common	04/03/1800	06/03/1800
John Deegin	Hampshire	Botley	04/03/1800	10/03/1800
James Weldon	Lancashire	Barlow Street, Collyhurst	25/03/1800	19/04/1800
James Austin	Kent	Bedgebury?	Unknown	02/08/1801
John Massey	Leics	Congerstone Heath	18/03/1801	23/03/1801
David Dutfield	Pembs	Unknown	Unknown	Unknown
John Palmer	Warwicks	Unknown	23/03/1801	01/04/1801
John Gubby and Jonathan Harben	Hampshire	East Parley Common	06/03/1804	12/03/1804
Thomas Temporel/Otter	Lincolnshire	Drinsey Nook	08/03/1806	14/03/1806
William Hove	Staffs	Gibbet Wood	11/03/1813	18/03/1813
Anthony Lingard	Derbyshire	Wardlow	22/03/1805	28/03/1815
John Rolfe	Cambs-Ely	Littleport Turnpike	Unknown	24/02/1823
William Jobling	Durham	Jarrow Slake	01/03/1832	03/08/1832
James Cook	Leics	Junction of Saffron Lane and Aylestone Road	04/08/1832	10/08/1832

Appendix 2: Maps, 1752–1834

These maps show the distribution of hanging in chains in England and Wales during the period of the Murder Act, between 1752 and 1834. Maps were compiled by using a base map created by D-maps (http://d-maps.com/carte.php?num_car=5583&lang=en). Where two gibbetings happened close together, the dots on the maps merge; thus, especially for London, a single dot may represent several events. Appendix 1 lists the individual events in full. The periods represented in each map are Map1a: 1752–1760; Map 1b: 1761–1770; Map 1c: 1771–1780; Map 1d: 1781–1790; Map 1e: 1791–1800; Map 1f: 1801–1810; Map 1g: 1811–1820; Map 1h: 1821–1830; Map 1i: 1831–1834.

136 APPENDIX 2: MAPS, 1752–1834

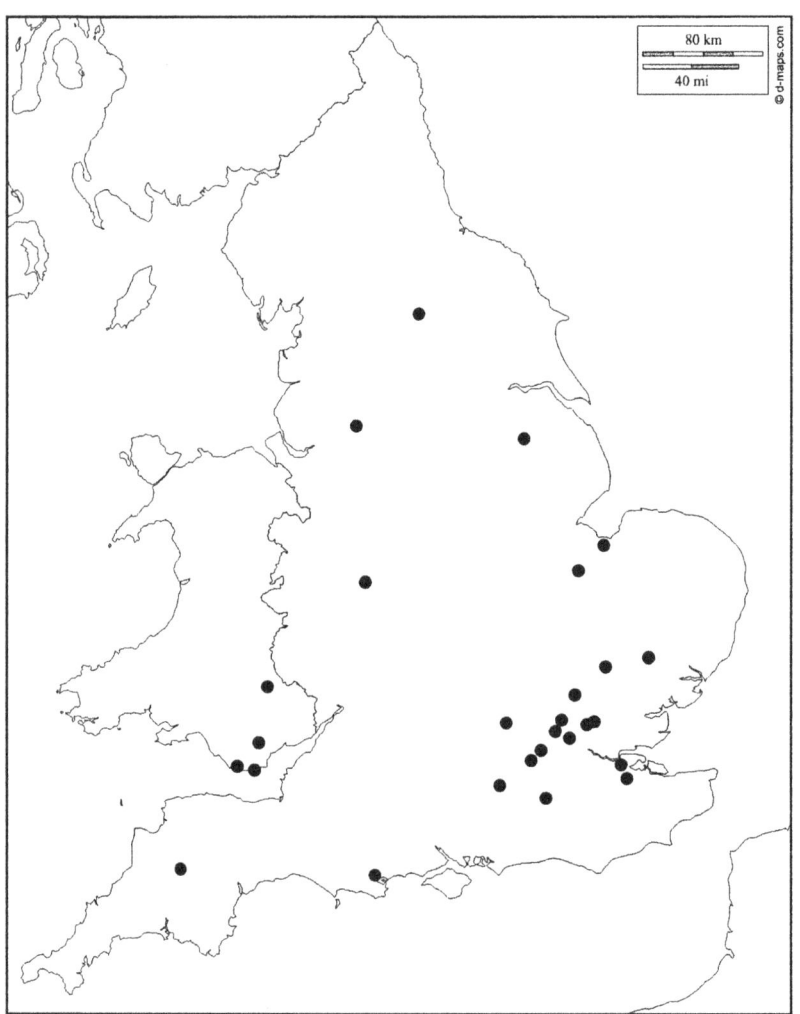

Map 1a   1752–1760

APPENDIX 2: MAPS, 1752–1834 137

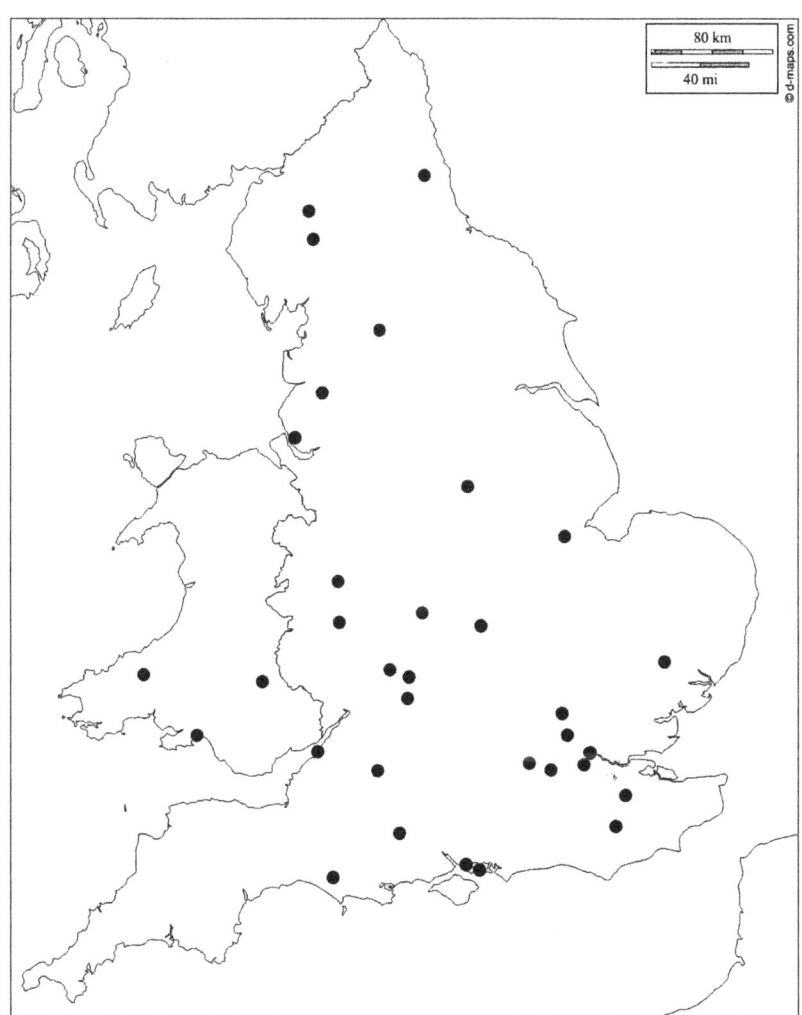

Map 1b 1761–1770

138 APPENDIX 2: MAPS, 1752–1834

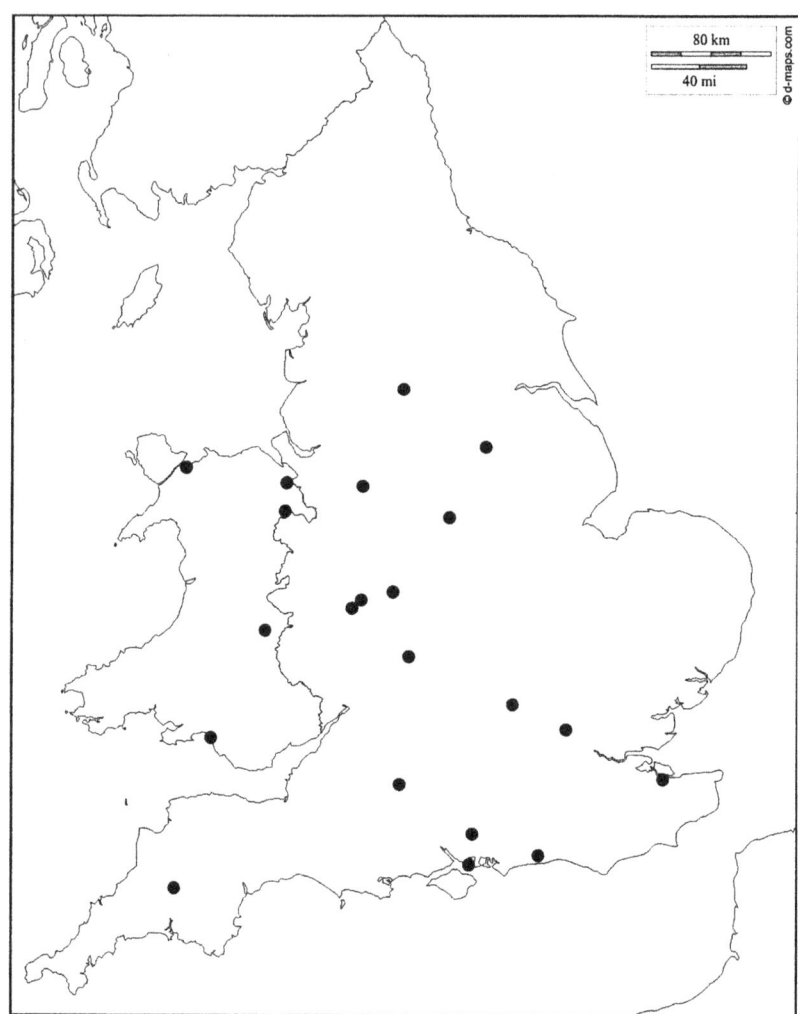

Map 1c 1771–1780

APPENDIX 2: MAPS, 1752–1834 139

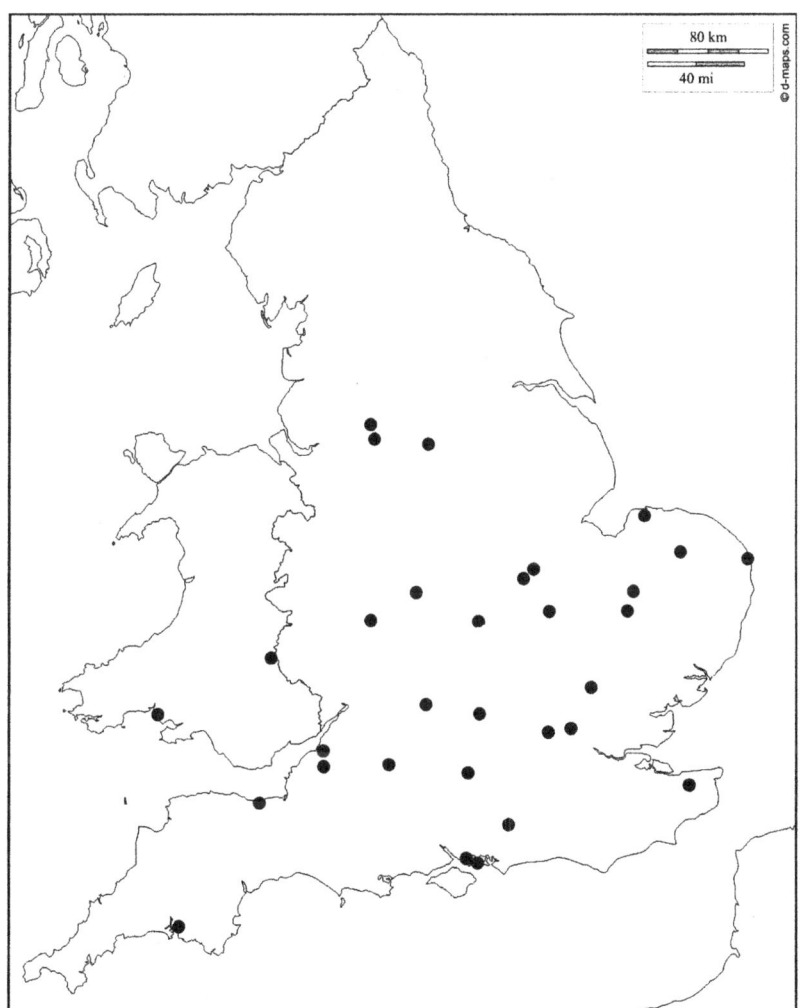

Map 1d 1781–1790

140 APPENDIX 2: MAPS, 1752–1834

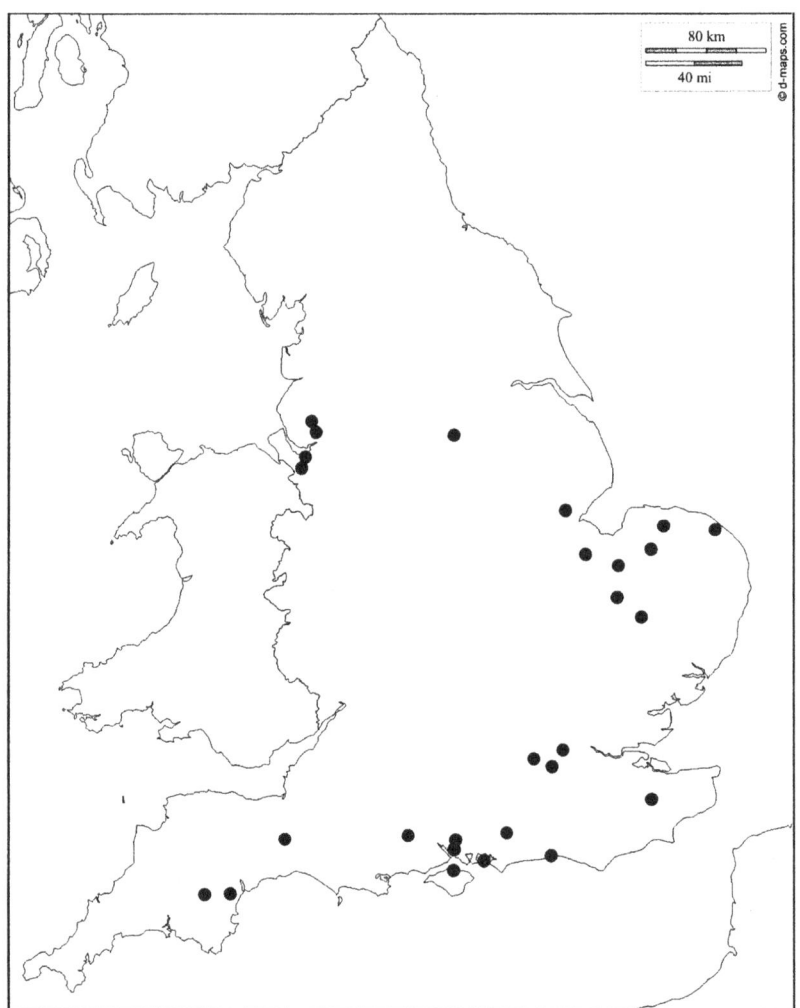

Map 1e   1791–1800

APPENDIX 2: MAPS, 1752–1834 141

Map 1f 1801–1810

142 APPENDIX 2: MAPS, 1752–1834

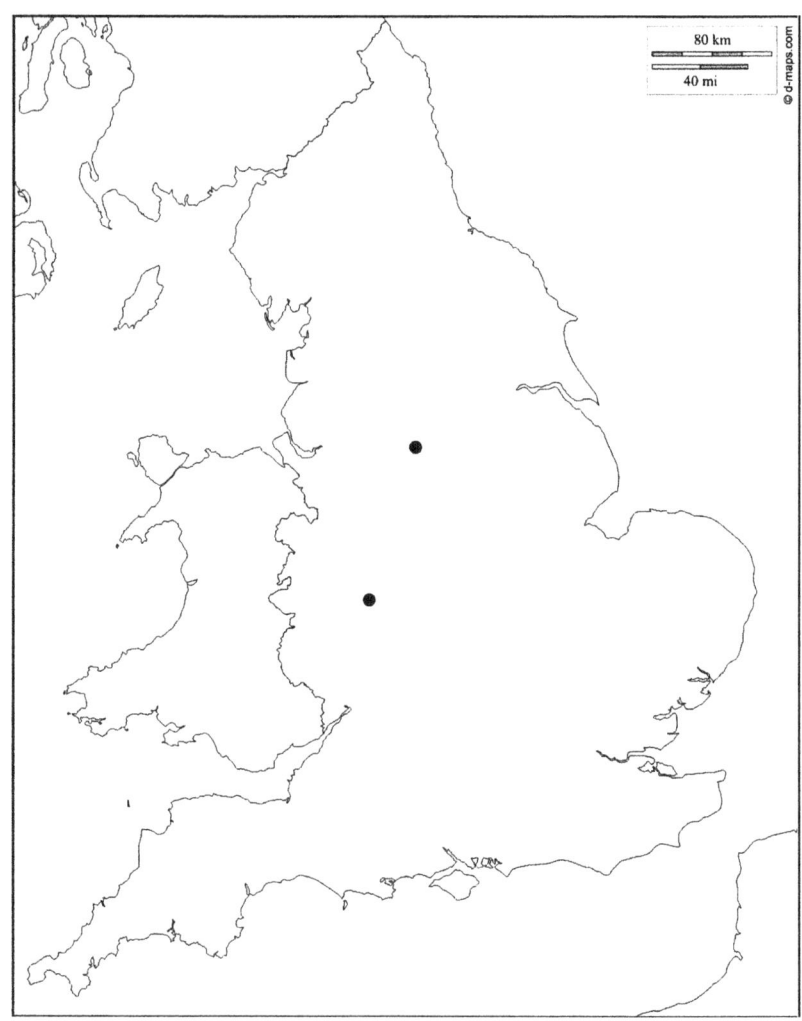

Map 1g 1811–1820

APPENDIX 2: MAPS, 1752–1834   143

Map 1h   1821–1830

144 APPENDIX 2: MAPS, 1752–1834

Map 1i   1831–1834

Concept Index

A
Admiralty courts, 10, 21, 27, 31, 49, 50, 61
Anatomisation, 9, 25
Anatomy act, 9, 28, 104
Assize courts, 10, 25, 32, 36, 49, 60

B
Body parts, 5, 6, 12
 hand, 77, 78
 head or skull, 5, 26, 79, 80, 92, 94–97
 stolen from gibbet, 59, 69, 79, 84
Burial, 10, 14, 16–18, 20, 28, 36, 85, 104, 106, 112

C
Crimes
 desertion, 14
 highway robbery, 6, 8, 10, 22, 108
 mail robbery, 22, 29, 108
 murder, 5, 6, 8, 11, 30, 34, 46, 50–59, 97, 108, 114, 115
 piracy, 10, 21, 29, 50, 108
 smuggling, 21, 23, 29, 60
 theft, 50, 69
 treason, 13–15
Crowds, 4, 29, 31, 34, 39, 43, 46, 50, 59, 74, 80, 82, 112, 114

D
Dead letter, 25, 26
Decomposition, 35, 39, 93
Dissection. *See* Anatomisation

E
Execution, 8, 9, 11, 15, 25, 31, 33, 34, 43, 45, 47, 62, 67, 75, 78, 80, 101, 106, 117
 beheading, 15, 16
 hanging, 15, 20; drawn and quartered, 11, 12, 14

F
Foucault, Michel, 15, 103, 104, 115

G
Gallows. *See* Execution
Ghosts, 4, 18, 31
Gibbet
 appearance, 77, 107
 cage, 31, 33, 35, 61, 64, 66, 67, 72–74
 placement/location, 21, 31, 33, 39, 43, 44, 77, 79, 82, 91, 109
 re-use, 40, 92
 structure, 27, 31, 47, 62, 66, 82, 85, 109
"Gypsies", 3, 85, 90

H
Hanging. *See* Execution

J
Jacobite rebellion, 12

K
Knobstick marriage, 2

L
Landscape, 5, 21, 46, 50, 111
Liminality, 60

M
Murder Act, 6, 8, 9, 11, 12, 21–23, 25, 27, 36, 102, 103, 105, 108, 117

P
Phrenology, 95–97
Post-mortem punishment, 11, 17, 20, 23, 36, 104, 106, 112, 117
 decapitation, 12, 14
 dismemberment, 15
 profane burial (Staking at a cross-road), 17, 19

S
Sheriffs' cravings, 5, 18, 34, 43, 61, 63, 68, 69, 73, 101
State power, 44, 46, 103, 104, 115, 116
 punishment and social control, 21, 23, 102, 103, 105, 106, 117
Suicide, 16–19, 36, 106

T
Tarring, 34, 35, 67
Torture. *See* Pre-mortem punishment

W
Wordsworth, William, 48

Historical Publications Index

B
Britannia, 46
Bury and Norwich Post, 26

C
Cumberland Pacquet, 17

D
Daily Courant, 72

G
General Evening Post, 76
Gentleman's Magazine, 18

H
Hampshire Chronicle, 13

L
Leicester and Nottingham Journal, 29, 80, 84
Lincoln Times, 4

Literary Gazette, 97
London Chronicle, 84
London Evening Post, 64

M
Morning Chronicle, 83
Morning Herald, 83
Morning Post and Advertiser, 83

O
Old Englander, 31
Oracle and Daily Advertiser, 83

P
Public Advertiser, 76

R
Read's Weekly Journal, 107

S
Sheffield Iris, 77

T
Times, 89

W
Whitehall Evening Post/London Intelligencer, 83, 85

Name Index

A
Abershaw, Jeremiah, 43
Abseny, Joseph, 107
Anderson, David, 69
Aram, Eugene, 49, 79, 89, 92–95, 97
Armstrong, Joseph, 36
Arsine, Francis, 68

B
Barchard, John, 64, 67
Benstead, Roger, 25, 74
Birch, William, 36
Blakemore, Ann, 27
Bocking, Hannah, 77, 108
Bowland, John, 73
Bradshaw, John, 15
Breeds, John, 69, 80
Broughton, Spence, 75, 77, 89
Brown, Thomas, 79
Burnet, Andrew, 83
Butlin, Richard, 24

C
Carleton, Robert, 34
Cato Street conspirators, 14
Chevin Highwaymen, 35
Clark, Daniel, 93
Clark, Jane and Eleanor, 26
Cliffen, James, 86
Cocklane, Matthew, 87
Colley, Thomas, 63
Conoway, Peter, 76
Cook, James, 28, 29, 34, 35, 66, 80, 104
Coombes, George, 84
Corbett, Edward, 24, 35, 47
Cromwell, Oliver, 15
Croxford, John, 24, 84
Cundell, William, 14
Curtis, John, 67

D
Deacon, Benjamin, 24
Deacon, Thomas, 12

NAME INDEX

De La Motte, Francis Henry, 13
Despard conspirators, 14
Drewitt, Robert and William, 74, 77
Dunsden, Thomas and Henry, 84

E
Evan, William Walter, 24

F
Fairall, William, 69
Felton, John, 40
Fletcher, George, 12
Flood, Matthew, 85
Franklin, John, 22

G
Gardelle, Theodore, 107
Gatward, John, 91, 107
Gill, Toby, 27
Gow, John and Williams, James, 59
Graham, Adam, 67
Grant, Jane, 77, 108
Grindrod, John, 87

H
Haggard, Rider, 67
Haines, John, 39, 83
Hammersley, Miss, 81
Hammond, John, 82
Hanks, Thomas, 23
Hawkhurst gang, 60
Hawkins, William, 63, 66, 82
Hazlitt, Robert, 76
Hooton, Philip, 84
Houseman, Richard, 93
Hurlock, Samuel, 81

Hutchinson, 94, 95

I
Ireton, Henry, 15

J
Jackson, Thomas, 45
Jarvis, Edwin George, 3, 85, 86
Jeffryes, Elizabeth, 26
Jobling, William, 28, 80, 104

K
Keals, John, 69
Kidson, Walter, 84
Kirkham, Mary, 2–4
Knott, John, 35

L
Leslie, Robert, 15
Lingard, Anthony, 76, 83, 91, 108
Lingard, John, 76
Lowry, Captain James, 59, 85

M
Mahoney, Matthew, 91
Matcham, Gervase, 48, 66
May, James, 24
Miller, Thomas, 4
Mills, Andrew, 90
Morey, Michael, 41, 92
Morgan, Thomas David, 12

N
Naden, John, 40

NAME INDEX

Nichols, John and Nathan, 25, 40
Nicholson, Thomas, 48, 62

O

O'Coigley, James, 14
Odell, William, 85
Otley, Thomas, 24
Otter, Tom, 9, 46, 85, 86, 90

P

Payne, Henry, 83
Peck, Joseph, 92
Pentrich revolt leaders, 14
Pitmore, Thomas, 82
Powell, William, 24
Price, James, 79, 86
Prothero, Jenkin William, 82
Pycraft, John, 26

R

Rawlinson, Martha, 2
Reading, Lambert, 107
Richardson, Michael, 76

S

Saxby, Robert, 26
Shuttleworth, William and John, 2
Smith, Gill, 84
Smith, John, 14
Smith, Ralph, 63, 90
Smith, William, 75
Spencer, John, 47, 83

Spiggott, William, 24
Stretton, John, 86
Sturabout, John, 69
Suffolk, William, 24, 86
Swan, John, 26

T

Temporel, Thomas. *See* Otter, Tom
Theobald, Jeremiah, 24
Tomkins, Gabriel, 86
Towneley, Francis, 12
Tull, Abraham, 63, 82
Tyrie, David, 13

W

Walton, Stephen, 67
Watkins, Thomas, 81, 82
Watson, Stephen, 86
Weldon brothers, 45
Wheldon, James, 107
Whitfield, John, 87
Whittle, William, 39
Willdey, Thomas, 44
Williams, John, 19, 20
Willot, Thomas, 68
Winter, William, 26

Y

y Gof, Sion, 69
York, 12, 89
Young, John, 43

Place Index

B
Bedfordshire, 35
    Dunstable, 86
    Luton, 35
Berkshire, 63
    Reading, 66
Bristol, 81–83, 85, 107
Buckinghamshire, 24, 47
    Bierton, 47

C
Cambridgeshire
    Huntingdon, 48
    Wisbech, 92
Cheshire
    Trafford, 79, 86
Cornwall, 36
    St. Ives, 66
Cumberland, 62
Cumbria, 48, 87
    Armathwaite, 87
    Brampton, 12
    Carlisle, 12, 67, 87
    Penrith, 12, 48

D
Derbyshire, 35, 40, 76, 101, 108
    Belper, 35
    Derby, 85, 87
    Pentrich, 14
Devon, 43, 62
Durham, 36, 90

E
East Anglia, 23
Essex, 36
    Chelmsford, 107

G
Gloucestershire, 24, 36, 45, 84, 91
    Avonmouth, 91
    Cheltenham, 36
    Tewkesbury, 36

H
Hampshire, 13, 36, 68
    Portsmouth, 40

© The Editor(s) (if applicable) and The Author(s) 2017
S. Tarlow, *The Golden and Ghoulish Age of the Gibbet in Britain*,
Palgrave Historical Studies in the Criminal Corpse and its Afterlife,
DOI 10.1057/978-1-137-60089-9

Herefordshire
　Hereford, 24
Hertfordshire, 63
　Puckeridge, 91

I
Ireland, 10, 12, 30
　County Louth, 29, 66, 69
　Galway City, 91
Isle of Wight, 41, 92

K
Kent, 14, 23, 60, 69

L
Lancashire, 87
　Pendleton Moor, 87
Lancaster, 17, 39
　Lancaster Moor, 3
Leicestershire, 101
　Leicester, 29, 66, 80, 102
Lincolnshire, 4, 28, 46, 63, 68, 84, 85
　Brigg, 27
　Drinsey Nook, 2, 4
　Lincoln, 102
　Saxilby Moor, 3
　South Hykeham, 2
　Surfleet, 84
London, 12, 19, 32, 36, 39, 49, 59, 60, 75, 76, 84, 108, 109
　Bow Common, 49, 76
　Cannon Street, 20
　Cato Street, 14
　Coldbath Prison, 19
　Ealing Common, 85
　Edgeware Road, 49
　Execution Dock, 59, 62, 84
　Finchley Common, 49, 75, 85, 86

Hounslow Heath, 40, 49, 83, 85, 107, 108
Kennington Common, 43, 84
North Heath, 77
Putney Heath, 43
Shepherd's Bush, 49
Stamford Hill, 81
St. Pancras, 12
Temple Bar, 12
Tottenham, 81
Wimbledon Common, 108
Windsor, 81

M
Manchester, 13
Middlesex. *See* London

N
Norfolk, 24, 26, 34, 43, 46, 49, 62, 67, 86, 87
　Bradenham, 73, 86
　Diss, 34
　Great Yarmouth, 96
　King's Lynn, 96
　Methwold, 45
　North Walsham, 86
Northamptonshire, 24, 101
Northumberland
　Elsdon, 26
　Whitley Bay, 47
Nottinghamshire, 2, 83
　Hockerton, 2
　Nottingham, 102
　Scrooby, 47, 83
　Treswell, 2

O
Oxfordshire
　Banbury, 47

R
Rutland, 45, 101
  Hambleton, 45

S
Scotland, 10, 12, 105
  Glasgow, 14
  Stirling, 14
Shropshire, 34
Somerset, 45, 61
South Yorkshire
  Attercliffe, 75, 77, 89
  Sheffield, 75, 89
Staffordshire, 40, 68, 72
Suffolk, 24
  Blythburg, 27
  Bury St. Edmunds, 41, 116
  Honington, 41
  Undley, 74

Surrey, 26, 45, 68
Sussex, 36
  Rye, 69, 80

W
Wales, 27
  Dylife, 69
Warwickshire, 101
  Birmingham, 72, 82
  Coventry, 44, 102
Wiltshire, 67

Y
Yorkshire
  Knaresborough, 49, 89, 92–94
  Malton, 96
  York, 12

© The Editor(s) (if applicable) and The Author(s) 2017. This book is an open access publication.

Open Access This book is licensed under the terms of the Creative Commons Attribution 4.0 International License (http://creativecommons.org/licenses/by/4.0/), which permits use, sharing, adaptation, distribution and reproduction in any medium or format, as long as you give appropriate credit to the original author(s) and the source, provide a link to the Creative Commons license and indicate if changes were made.

The images or other third party material in this book are included in the book's Creative Commons license, unless indicated otherwise in a credit line to the material. If material is not included in the book's Creative Commons license and your intended use is not permitted by statutory regulation or exceeds the permitted use, you will need to obtain permission directly from the copyright holder.

The manufacturer's authorised representative in the EU is Springer Nature Customer Service Centre GmbH, Europaplatz 3, 69115 Heidelberg, Germany. If you have any concerns regarding our products, please contact ProductSafety@springernature.com

Printed and bound by CPI Group (UK) Ltd, Croydon, CR0 4YY

23/03/2026

02076355-0010